How to Make Money by Keeping Ducks
also, The breeding and management of the most useful varieties of geese

by Henry Huddersfield Digby

with an introduction by Jackson Chambers

This work contains material that was originally published in 1897.

This publication is within the Public Domain.

This edition is reprinted for educational purposes
and in accordance with all applicable Federal Laws.

Introduction Copyright 2018 by Jackson Chambers

The World's Largest Selection of Vintage Poultry Books

www.VintagePoultry.com

Self Reliance Books

Get more historic titles on animal and stock breeding, gardening and old fashioned skills by visiting us at:

http://selfreliancebooks.blogspot.com/

Introduction

I am pleased to present yet another title on Poultry.

The work is in the Public Domain and is re-printed here in accordance with Federal Laws.

As with all reprinted books of this age that are intended to perfectly reproduce the original edition, considerable pains and effort had to be undertaken to correct fading and sometimes outright damage to existing proofs of this title. At times, this task is quite monumental, requiring an almost total "rebuilding" of some pages from digital proofs of multiple copies. Despite this, imperfections still sometimes exist in the final proof and may detract from the visual appearance of the text.

I hope you enjoy reading this book as much as I enjoyed making it available to readers again.

Jackson Chambers

PREFACE.

Upon the completion of the original manuscript of this unpretentious work I voiced to a friend certain fears I entertained as to its reception by the public generally, and the fancying section of the public in particular. His response was in the nature of an assurance that I could stand any criticism which might be levelled at me through it.

Granted that I am not averse to criticism, I must, nevertheless, express the hope that my readers will deal leniently with me, not because my views as herein set forth will not bear criticism, but because in all I have written I have conscientiously striven to say only those things which ripe experience teaches are the best that can be said on the subject.

The primary object I have in view in compiling the treatise—as the title conveys—is to show "How to make £50 a year by keeping ducks," and also how to augment that income by breeding geese. I have, consequently, dealt only in my work with those varieties of ducks which I consider best fitted to accomplish the object to be achieved.

The breeding and rearing of ducks, both for exhibition and market, are minutely treated of, the reader being put into possession of all that it is necessary to know in order with ease and confidence to put the theory I propound into successful practice.

PREFACE.

Up to the time of writing this book no attempt had been made to draw up standards of perfection for waterfowl; on the ground that the birds were generally judged by weight. Admittedly, size, *i.e.*, weight, is an important point in the most useful varieties of ducks and geese. At the same time there are many more equally important points to be considered.

The numerous enquiries I have had from beginners, concerning this matter, convince me that a scale of points would be beneficial, especially to young fanciers. In consequence of this I have drawn up standards of excellence for the varieties on which I have written, and, with the assistance of several members of the Waterfowl Club, I am here able to place before my readers recognised standards and scales of points, which I feel sure will be welcomed by, and be a boon to, waterfowl fanciers generally.

THE BURNE, 1891.

PREFACE TO THE FOURTH EDITION.

THIS issue represents the Fourth Edition of my book. The success with which my humble literary endeavour has from the onset been attended has far outmeasured my fondest expectations. For thus extending to me an unqualified meed of kindness and support I am constrained to offer to the public and the Press alike my sincere and hearty thanks.

That I was justified in introducing standards of perfection and giving scales of points has been shown by the adoption of such standards and scales at the annual general meetings of the Waterfowl Club, held in Liverpool in 1892 and 1897, and by a general consensus of opinion in favour of their universal recognition and use.

Since first publishing my work I have added to its scope in a manner calculated to increase its usefulness and, generally speaking, to enhance its value.

Amongst the more important additions is the chapter dealing with "Indian Runner" ducks. To this addition I perhaps naturally attach the greater importance, from the circumstance of the original object I had in view in compiling this book was the improvement of the waterfowl species generally. No doubt my readers will recognise in the standard of perfection appended to the section on the "Indian Runners" a something which supplies a long-felt want. This standard was adopted at the annual general meeting of the Waterfowl Club in January, 1897, and is here reproduced (all rights reserved) with the sanction of the said Club.

THE BURNE, 1897.

INTRODUCTION.

In introducing this little work to small farmers, gardeners, and cottagers, I do so with the desire to encourage the breeding of *first class* stock, not necessarily for exhibition, but also for table purposes.

At the outset let me lay down what I consider to be a fundamental principle, viz., that for successful breeding for exhibition and marketable purposes alike a good class of bird is indispensable to begin with. I shall try to place before my readers that which I consider practical and easy of achievement, as it is far from my purpose to write from a supposititious standpoint. Herein will be found plain rules for the successful management of ducks and geese, and also tables of the different points and their relative show values in five breeds of ducks and two of geese, thus enabling the farmer or young fancier to assess the excellence or otherwise of his stock.

It has been my aim to show that profit can be made by farmers, or in fact anyone with convenience to keep ducks and geese, if endowed with patience and perseverance. To mention one use to which ducks and geese naturally adapt themselves, and which is sometimes overlooked, I may allude to their value amongst vegetable crops, which they very eagerly free from the injurious influence of caterpillars and other insects. Mr. James Byne, of Woodbine Cottage, Harristown, County Kilkenny, writes to the *Weekly Freeman* on the subject of "Flukes" in sheep, and after an able description of the cause, proceeds to give the remedy as follows: "My family, always fond of poultry breeding, added to their large stock by importing some prize breeds of ducks and geese, which were kept on the low-lying parts of the farm. Strange to say, soon after the introduction, the 'Flukes' finally disappeared from the land, owing, I am certain, to the ducks and geese feeding on the slugs. I would strongly advise any farmer suffering from 'Flukes' to try this simple remedy." The foregoing evidence ought to be sufficient to prove the value of ducks and geese kept on farms and in large gardens, if only for the sake of cleansing the land of objectionable and destructive insects.

The prices obtainable for ducks and geese as marketable commodities are far in excess of what they were in the earlier years of the century. Experience shows that superior specimens are always the most profitable to breed and rear, even for table purposes.

It is estimated that London pays upwards of £30,000 per annum to the town and villages surrounding Aylesbury for ducklings. I know for a fact that it is not at all uncommon for a ton weight of ducklings to be sent from Aylesbury to London in one night during the season. These are killed, plucked, and packed in flat hampers. They are collected by the railway companies and forwarded direct to the salesmen, who, I am informed, are very prompt in making their returns.

Now, if Aylesbury and the villages around can rear and send £30,000 worth of ducklings to the London market, I should like to ask our farmers, gardeners, and cottagers if it is not possible for those resident within easy reach of our Northern towns to breed a proportionate quantity of ducklings and sell them direct to the consumers?

There are hundreds of people living sufficiently near some of our large towns who could add considerably to their present income by keeping and breeding ducks and geese for market purposes alone, and more especially if they keep pure-bred stock and reserve a few of the very best for exhibition or stock purposes. It is an admitted fact that our British ducks and geese stand pre-eminent. When we remember the prices annually obtained for early ducklings and Christmas geese, it is demonstrative evidence that there is a good demand, especially for *first-class* articles of food of this description.

HOUSES FOR BREEDING DUCKS.

WATERFOWL should always have a lodging place of their own. They should be kept apart from other fowls, and under no circumstances be allowed to sleep beneath their perches. Still, where only a very few fowls are kept, a little arrangement could be made which would make them a comfortable berth in a hen-house. This could be done by placing loose boards under all the perches, so that the excrement from the fowls would fall upon them, and so protect the ducks and geese from what would otherwise be very objectionable. The better plan, however, is to house waterfowls in separate dormitories They can then be kept very much more comfortable and free from vermin.

A duck-house can be made out of almost anything. varying from a stall in a stable or cow-house, or any unoccupied out-building frequently seen about a farmhouse, to a pigsty or the lower part of a hen-house, often lying dormant behind a cottage. I have seen a most comfortable dwelling for ducks made under a raintub. In this case there was a square basement built of bricks, 4ft. by 4ft. and 4ft. high, with a door in the front 3 ft. 6 in. high by 2 ft. 6 in. The floor was set with bricks. There was a small square of glass and a ventilator in the door.

On more than one occasion I have come across a waterbarrel itself converted into an abode for ducks. A very large cask or barrel turned on its side will answer this purpose.

About three battens cut to fit the bottom of the cask and a few boards 1 in. thick make a capital false bottom, which can easily be taken out for a thorough cleaning.

One end of the barrel is taken out and a door made about two feet wide, the whole height of the cask end. The material taken out of the end comes in for the door.

Bore a few holes in the top of the door for ventilation, cover the top of the barrel with any kind of felt, old oilcloth, canvas, or anything available, and then give it a good coat of tar, and you have a very nice house for two or three ducks and a drake at the cost of a few shillings.

I make a capital duck-house out of two piano cases, as shown in sketch No. 1, which can be bought in most

DUCK-HOUSE MADE FROM TWO PIANO CASES.

towns for five shillings each. Knock the backs completely out, take tops off, and then place the two cases back to back. Make the door in the centre of one end, taking either two or four boards out of the gable end. Nail 2 in. by 1 in. battens on the inside of these boards, and your door just fits the place out of which it came. All the additional wood required to complete one of the most useful and comfortable houses for a breeding pen of ducks is about 3 ft. 11 in by 1½ in. board. This board is sawn diagonally to form the centre of the ends, and is nailed on the top of the ends By this means you add 11 in. to the height of your cases and avoid sawing the ends to get a fall for the roof. The wood taken out of the backs of the two cases will be more than sufficient for battens for the door, also for two battens across the outside of the two cases and two across the inside of the ends. There will be sufficient for a false bottom if you care to put one in. If you do not wish to put a false bottom in your house the spare wood will come in to make a nice little shed by the side of the house, under which you can put your gravel, old mortar, etc. Don't forget a few holes at the top for ventilation. The boards being nailed on the roof crossways will, of course, require covering with felt, five square yards will be sufficient.

Your house being now really built, only requires a good coat of lime-wash inside and well tarring outside. A nice bunch of straw shaken on the floor and your house is ready for habitation.

If you get ordinary-sized cases the dimensions will be about 5 ft. by 4 ft. 6 in. and 5 ft. high, sufficiently large for a breeding pen of ducks. It will also be large enough to accommodate an average brood of ducklings. It is easy of access for the collection of eggs and cleaning out The whole cost of material, including hinges, staples and padlock, is fifteen shillings. This is not allowing for labour but even after making an allowance of five shillings on this account, we have a really good duck-house for £1.

I do not believe in large houses and large flocks of ducks, especially for stock purposes. My experience is that a small flock of ducks comfortably housed pays much better than a large flock all running together and housed in one large shed or room.

DUCK-HOUSE AND ENCLOSURE.

THE illustration on the following page is a sketch of one of the most useful houses for rearing a large quantity of ducklings for market, or a less number for exhibition, containing, as shown on the plan, eight large and commodious pens all under one roof, six of which are 9 ft. by 9 ft., and two 18 ft. by 9 ft. The partitions dividing the pens are 3 ft. high.

The enclosure is 27 ft. by 18 ft., and the water-tank in the centre represents a galvanized iron cistern, 6 ft. by 3 ft. 6 in. and 1 ft. deep. Water must be conveyed into this tank in the way most convenient to the ducker. My method is by means of a 1½ inch pipe, and is run off by means of a brass plug in the bottom, and a drain falling to the lowest level of the ground outside the structure.

The outside measurement of the building, which is of wood, is 36 feet × 36 feet, height at ridge 8 feet, and at eaves 6 feet, and the gates, which are double, are 7 feet wide. The doors are 5 feet × 2 feet 6 inches, over which and right along the top of the house, under the eaves, is run a piece of strong wire netting one foot wide, thus giving abundance of light and ventilation.

The floors may be of wood, brick, or concrete. I prefer the latter. The roof may be covered with felting corrugated iron, or thatched with straw, but I find that felting, well tarred, answers best.

As the doors of the pens open out into the enclosure, each lot of ducklings can be turned out to feed and water separately, during which time the pen can be thoroughly cleaned out.

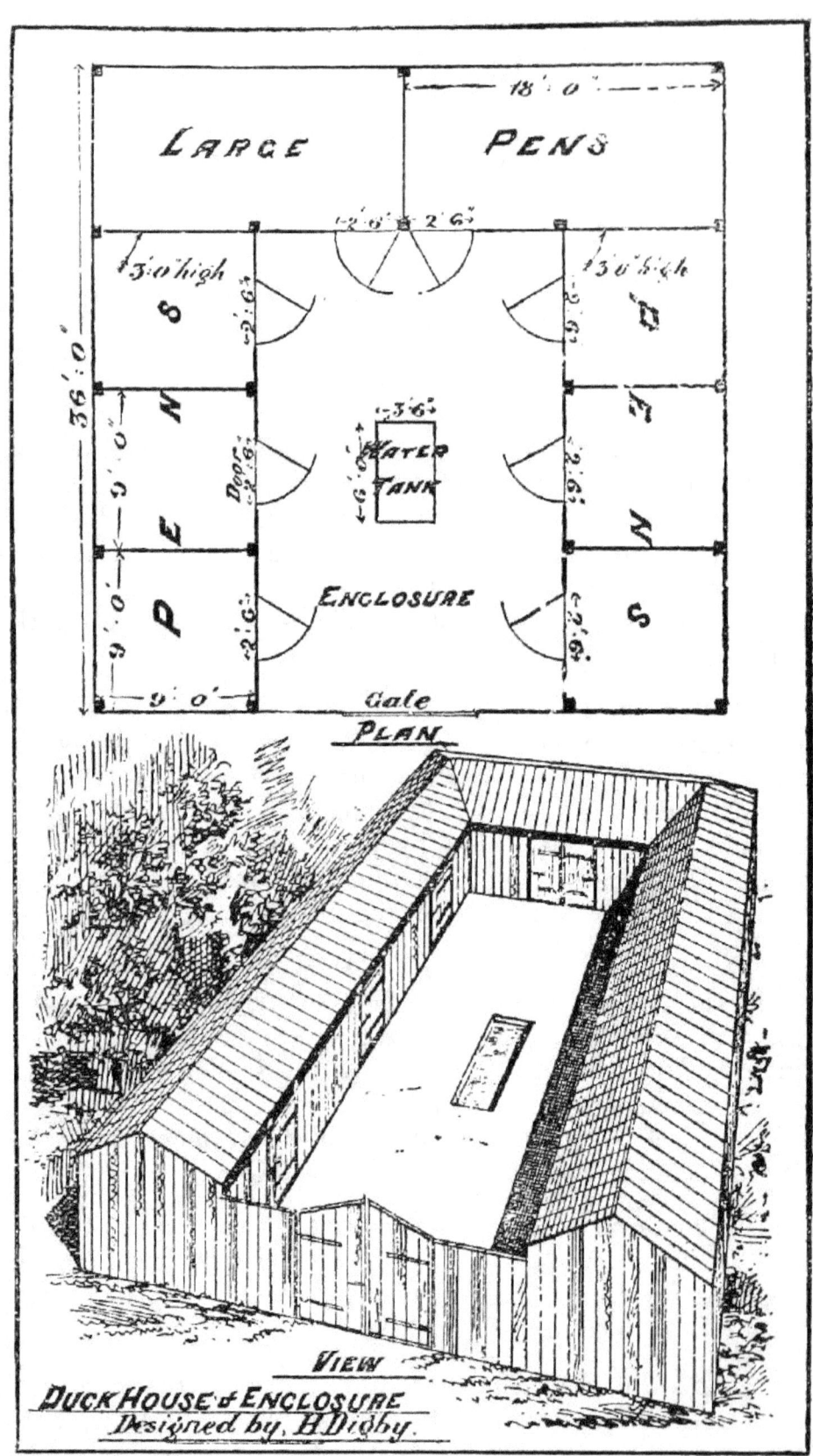

The advantages to the duck-keeper of a house of this kind are many. In the first place it is more suitable for rearing a large quantity of ducklings either for market or exhibition, or the two combined. It is also admirably adapted for housing stock ducks. It would also answer for geese and for preparing any kind of heavy waterfowl for exhibition.

Such a house can be put up for very little cost, especially if the ducker is a handy man with the saw, hammer, and nails, and can put it up himself.

This plan of mine has also another advantage, for should its dimensions be too large or too small for the purse, space, or requirements of my readers, they may add to or diminish the size according to their own convenience.

This house and enclosure is, according to my own idea, most practical and inexpensive, considering the amount of accomodation it affords.

Sketch No. 3 is a correct representation of the interior of my duck-house now in use at "The Burne," and although it only shows three pens on each side there is in reality twenty pens in this house, each pen being 6 ft. by 4 ft.—total length 40 ft. and 18 ft. wide; the passage in the centre being 6 ft. wide; the height at the ridge is 13 ft. and at eaves 7 ft. The pens are divided by wood partitions 3 ft. high and continued to the top of the house by a frame of wood laths. The doors are 6 ft. by 3 ft. 6 in. and are boarded and lathed same as the partitions; it is lighted by means of skylights, and ventilated at the top of each end, so that it is useful not only as a duck-house but also as a poultry-house. The floors are concrete, so that they can be easily cleaned out by swilling with water. The roof is covered with strong felting, the whole structure well tarred outside and lime-washed inside. It is a most convenient house, and was not expensive. It has the same advantages as No. 2, so far as size and requirements are concerned.

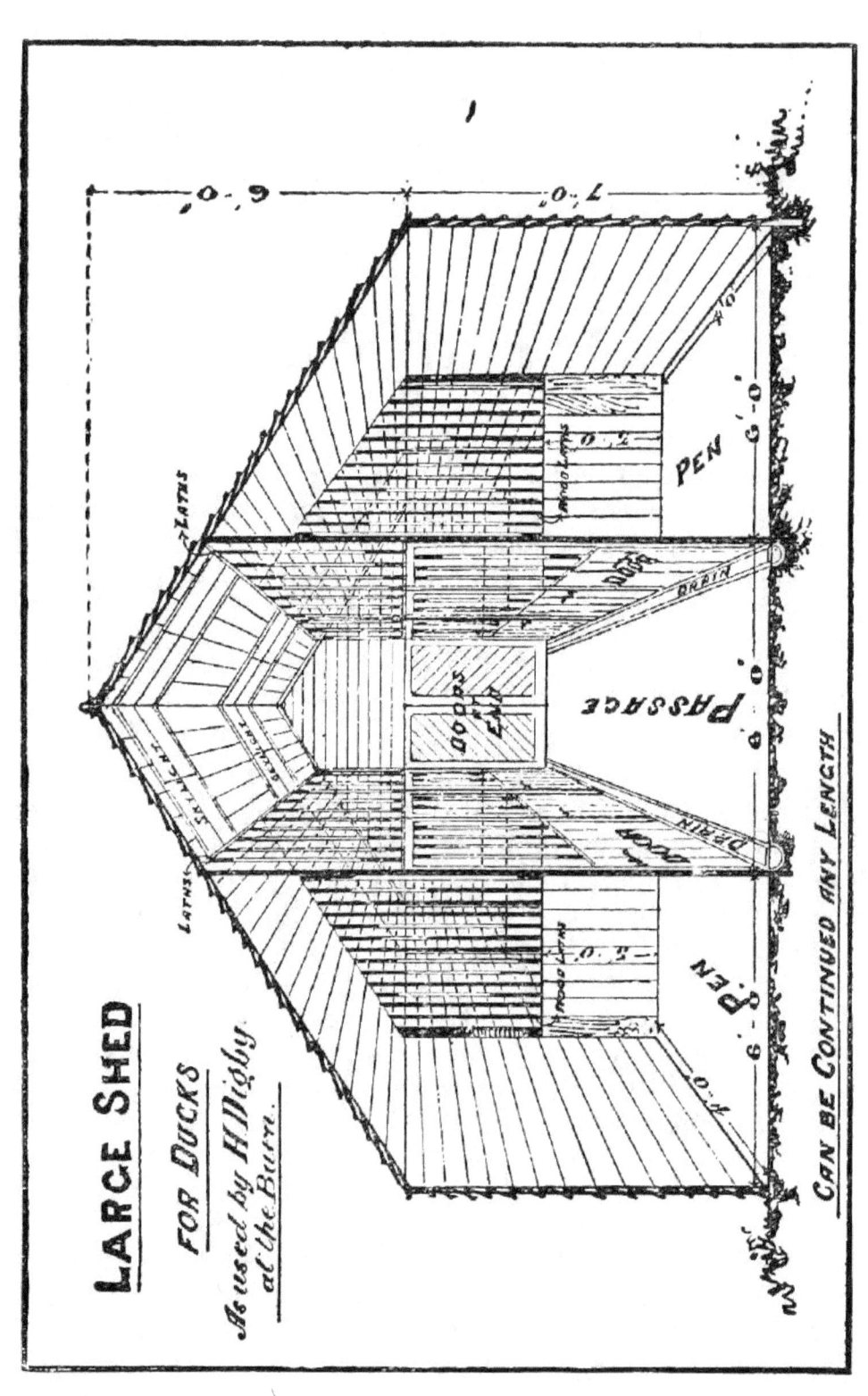

PONDS AND WATER.

One of the foremost considerations for those contemplating the breeding of ducks and geese is that of an abundant and constant supply of water. As their generic title—waterfowl—implies, their first instinct is in the direction of water. A first essential to successful breeding is ready and regular access to *water*. It does not necessarily follow that there should be a running stream, but a pond or pool of such dimensions as to permit of the birds taking a bath is an absolute necessity for the free propagation of their species. For rearing for exhibition purposes the supply of water for bathing must all along be unstinting and easily accessible. Whereas in rearing for the table a less extensive supply might suffice, nay, in fact, at certain seasons of the year is desirable.

Ponds for ducks are made all shapes and sizes, and of nearly all kinds of materials.

The first duck-pond I had was the end of a large oil-cask, about nine inches deep and three feet in diameter. This was sunk in the ground, so that the top was level with the surface, and when there was no water draining off the land, I had the pleasure of filling it with water out of a draw-well, which I had sunk close by. As time went on I grew tired of drawing the water out of the well with a pail, so had a pump fixed, which answered very well for a time, until I began breeding rather extensively, when I found it necessary to have a larger pond. To meet emergency, I bought a galvanized iron cistern, 6 ft. by 3 ft. 6 in. and 1 ft. deep. This was sunk in the ground in the same manner as the cask end above referred to. It had a brass plug on the bottom, so that the water could be let off through the drain beneath and be cleaned out. This cistern is one of the best ponds I ever used for ducks, it has been in use now for over twelve years, is still in sound condition, and at this date—October 28th, 1897—is still in use in one of my runs at "The Burne." During the twelve years use of this cistern it has been removed from one place to another on two occasions, and has never cost a single penny for repairs.

It is a very easy matter making a pond if you have a constant stream of water. Some of my ponds at "The Burne" are twenty yards long, and vary in width from one to six yards. In the first place we make a wall of stones or bricks, just leaving a space for a 6 in. iron pipe in the bottom This pipe is to run the water off to clean the pond out. Build a wall across the lower part of the stream, then build another wall of bricks about a foot distant from the first, and fill in the space between with clay made into puddle.

In flat countries all that is required is a little excavation where there is any run of water, and you have soon a duck-pond.

The difficulties in making suitable ponds are where there is little or no water. Then comes in the use of bricks, stone, wood, and cement, or better than all, a good galvanized iron cistern, which is not expensive, holds a good body of water, takes up very little room, and, by the adoption of a plug, can be cleaned out easier and better than any other pond.

A certain writer asserts that ducklings "will grow faster without water than with it," which in my opinion is a great mistake Water is the natural element for ducks, it is *not* in their nature to thrive without having access to it, therefore, if you want perfect symmetry, a bright eye, and a general bloom throughout, you *must* attend to nature's behests, and not allow your ducklings to die in the sun They will *not* die in the sun if they can get into water.

HOW TO BEGIN AND THE SELECTION OF STOCK.

I wish it to be distinctly understood that I am not presuming to show capitalists how to make £500 a year simply by multiplying my figures by ten. But I know by experience that there is money to be made by keeping ducks, and I will endeavour to show small farmers, gardeners, cottagers, and new fanciers, having a desire to keep, and still a fear of keeping, ducks, that they may do so successfully, and that without spending very much money at the commencement.

I often smile when reflecting on the course I took in beginning my career as a fancier. Always from my youth upwards imbued with a lofty ambition, I started out on the venture with the fixed intentions of making a bold bid for first-class honours from the onset. I sent my orders and cash to the best breeders in the kingdom and asked for the best birds that could be bought. No doubt I got what I asked for, which cost me a lot of money, yes, more than I ever dare tell my better half.

Then, like many more impatient novices, I entered my birds for *all* the Shows in the district, on some occasions winning the coveted premier honours. Not satisfied with reasonable success, I still went on exhibiting at every Show within easy reach until my once beautiful ducks were either all dead or nearly so

Experience proverbially teaches wisdom. After many necessitudes peculiar to beginners, and by dint of persistent observation I at last learned sufficient to enable me to make profit out of my hobby.

My first words to the fancier in embyro is, *be careful* in the purchase of your eggs or the selection of your breeding stock.

The selection of eggs or stock is one of the most important requisites for success in breeding ducks and geese.

It is most desirable and important to learn what variety of waterfowl is most suitable for your circumstances and accommodation. All ducks are not alike They differ considerably in size, colour, etc. Some varieties would thrive and do well under conditions which would prove fatal to others. The bills of the Aylesburys would undoubtedly be affected on a ferruginous soil, as also would the keeping of them under certain other disadvantages; whereas the Pekins would thrive and arrive at the highest state of perfection on such soil and under such circumstances; whilst the Rouen and Cayuga would flourish on a class of land and on waters which would be detrimental to either the Aylesbury or Pekins, so far as exhibition points are concerned. By the same rule the white Embden goose should not be selected for land and water containing much iron, for the white plumage would certainly be affected by it. Therefore I would advise beginners to consider well the class of ducks or geese best adapted to their circumstances.

Previous to purchasing your stock birds be sure that you have house or houses and pond complete. If you wish to be successful *keep one variety only*. Of course you will keep that variety which best pleases your taste, if adapted to your circumstances and accommodation. Personally, I have found the Aylesbury ducks most profitable, their reputation for table purposes and egg production being unrivalled. They are hardy, good foragers, and lay an abundant supply of large eggs. Their colour being pure white, there is no fear of foul feathers, bad pencilling, &c. Their progeny are, under ordinary circumstances, fast growers, and put on flesh at a very early age. I believe I am correct in saying that they arrive at maturity sooner than any other variety, and, for all-round purposes, I believe they are the best and most profitable of all ducks.

Next to the useful Aylesbury comes the magnificent Rouen. There is an excellent representation of a Drake of this variety in Mr. Lewis Wright's book, in fact it is the best I ever saw, but it is tame compared with the metallic hues and irridescence of the plumage of a living specimen in first-class condition. The grand colour and pencilling of a first-class Rouen duck are exceedingly pretty, but unless breeders are on their guard they will eventually ruin the constitution of this beautiful variety.

Size and stamina should be considered quite as much as colour and marking. Therefore let me caution breeders not to sacrifice size for the sake of a too fine marking. Do not select your stock birds too closely related. I know some strains are already degenerating, and if in-breeding is indulged in much further, size and stamina are sure to suffer.

The Pekins will live and do well in many places where the Aylesburys could not, and the fact of their being so very wild when at exhibitions may be accounted for by their enjoying perfect liberty at their homes. I have known Pekins sent to an exhibition without any preparation whatever, having been lifted off their runs and sent direct to the show, where they have secured first honours, being in splendid condition. This variety looks larger than it really is, being much denser in plumage, but does not come up to the Aylesburys or Rouens by some pounds in weight. They are, however, very good layers and foragers, and when crossed with the Aylesburys mature very early, and make capital birds for the table.

The Cayuga is a very handsome and useful variety, and deserves to be better known and encouraged, its brilliant green-black plumage and great size being sufficient to recommend it. The flavour of its flesh is considered by some people superior to any other of our domesticated varieties. I can strongly recommend it to gentlemen for their own tables. It is not only useful, but exceedingly ornamental, and suitable for any kind of land. There is no denying the fact that crossing improves the size and stamina of ducks. Still, I do not see the advantage of *crossing two different varieties*, even for market purposes. There are so many fanciers of any one variety nowadays that it is easy enough to get a drake which is not related to anyone's own ducks without introducing the blood of another variety into the veins of our much-prized pure-bred stock, which is just as good for the table as a mongrel would be. The very best would certainly be worth more than killing price. The few best should realize the bulk of the profit.

In making your selection of breeding-stock ducks be sure that the birds you are buying are comparatively young. If you purchase ducks two years old, let your drake be only one year old, or *vice versa*. Ducks arrive at maturity much earlier than geese. My experience is that the best results are obtained from birds of one or two years of age. On no account would I advise anyone to purchase stock-ducks over two years old or thereabouts, although they frequently breed well in their third year. Three or four ducks are sufficient to put to one drake. The ducks should be selected from one yard, and the drake from another.

Some extensive breeders of *one* variety might be able to supply both unrelated, but satisfy yourself that the stock you are buying is young, the marking rings introduced and adopted by the Waterfowl Club being a safeguard in this respect. Do not on any account be tempted to purchase old, used-up exhibition birds for stock purposes, for nothing but loss and disappointment would follow.

Buy your stock birds from reliable breeders, and, if possible, go and look at the whole stock you select from. You would then be able to judge who are the breeders of the best birds. Be aware of misleading advertisements, for there are a few such put in some of our poultry journals occasionally. Don't put too much confidence in wholesale dealers and professed breeders of numerous varieties. They are seldom able to win at a good show, even with their very best. If they cannot win with their very best, you may rest assured that they cannot sell birds for others to win with, and there is little or no dependence to be placed on their stock. In consequence of rapid and continued changes being made in their stock, it never becomes a strain in which there is any certainty. These dealers are continually buying and selling, just as opportunity presents itself, consequently their birds cannot be relied upon for stock purposes in the same degree as can birds from the stock of a skilful, successful, and careful breeder only.

My opinion on the way to success is to make a proper start. Go to some breeder of the one particular variety in which you have decided to invest your money, and there make your selection according to your requirements and other circumstances. Do not on any account be tempted to purchase old birds. Invest your capital and energies in young, healthy stock. In perusing one of our poultry journals a short time since, I noticed an article written by one of these wholesale dealers, in which he cautions beginners not to go to large exhibitors for stock birds, on the ground that they often buy their exhibition birds from other breeders. This may be an exception, but it is certainly not the rule. If new beginners will take my advice they will go to the large breeders and exhibitors for their stock birds. Such large breeders and exhibitors are always prepared to supply stock birds at reasonable charges, and in very many cases at sums only a little over killing price. In fact it is from these large breeders and exhibitors that the wholesale dealers get the bulk of their merchandise. A pamphlet lies before me as I write in which are offered eggs for sitting from no less than 80 separate pens of poultry, ducks, geese, and turkeys, together with nearly a thousand and one other articles varying in form from poultry houses to poultry pills and insect powder He must have a very large establishment, also an immense staff of men or pupils. I fancy the staff will consist of the latter, as the fee for three months' tuition is something over £10. Now, according to the above list, this gentleman must have 80 separate pens of stock birds from which to supply eggs. There appears to be no allowance made for more than one pen of each variety. Consequently, the vendor expects to breed both cockerels and pullets up to the present state of perfection from one pen of stock birds!

I will not dwell longer on this point, further than ask two questions. First,—Does it not take ten separate breeding pens of Hamburghs to produce five pairs of exhibition birds? Secondly,—Does it not take two breeding pens of ducks to produce one pair of exhibition Rouens?

It will be to the advantage of all beginners if they take time and carefully consider the importance of the selection of stock.

Pair of Modern Aylesburys. The Duck winner of three first prizes and two cups at the Crystal Palace; also, three first prizes, two cups, and the challenge cup at Liverpool, 1889. Drake winner of the challenge cup at Liverpool, 1890, &c. Bred and exhibited by the Author of this work.

MODERN AYLESBURY DUCKS.

This famous breed of ducks, being my own speciality, naturally occupies the first place in my estimation. The great size to which the birds attain at an early age, the excellent flavour of their flesh, and the tempting appearance they present when killed and nicely dressed, are qualities all unequalled by any other breed. They are very prolific, hardy, and thorough foragers; they thrive in almost any climate or on any soil, doing equally well in the South of England or in the North of Scotland; they also do well in Australia and America. I have sent several consignments to these countries during the last ten years, and have frequently heard good reports from my customers. Their name suggests that their centre is Aylesbury, in Buckinghamshire.

No doubt there was a time when there were reasonable grounds for supposing that the Aylesbury duck was a local variety. In fact, it was generally thought (not more than 25 years ago) that this breed could only be kept in perfect condition in the Vale of Aylesbury, but as the number of Poultry Shows increased, and with them fanciers, the Lancashire fanciers soon began to share the honours with exhibitors hailing from Aylesbury.

A little later on followed a well-known Scottish fancier, who also managed to keep the Aylesburys in fine condition. Then came Mrs. Stinton, of Darlington, who bred and showed the Aylesbury duck in faultless condition. After this came a Yorkshire fancier, who succeeded in breeding and exhibiting the Ten Guinea Challenge Cup winner three years in succession, thus winning the trophy outright. This Yorkshireman, who is none other than your humble servant, flatters himself with the knowledge of the fact that he is able to keep Aylesburys in perfect health and condition, although his land is very heavy and damp, the sub-soil being yellow clay. I am, however, bound to admit that in the "County of many Acres," and the North generally, we cannot breed these birds so early in the season as they can in the South of England.

For home consumption and market purposes let me strongly recommend this breed to farmers and others, for they excel in every particular. In the first place, they arrive at maturity sooner than any other variety of ducks. They are as hardy as crows, attain to a great weight, and are wonderfully prolific. The London markets are regularly supplied with enormous quantities of them during the season.

Duck-breeding is carried on in and about Aylesbury by what is termed "a superior class of labourers." These are men who have saved up money enough to secure an immunity from hard work, and who do not grudge giving their whole time and attention to this pursuit. During the months of November and December high prices are paid for Aylesbury duck eggs for sitting, ten shillings and twelve shillings per dozen being no uncommon price at that time of the year, the purchaser running the risk of their proving fertile.

The streams and ponds within several miles of Aylesbury are looked upon as common property. Hundreds of white ducks, with patches of paint of various colours dabbed upon some particular part of their plumage, may be seen on the river Thames, or in the brooks, or on the ponds. These distinctive marks are the marks of ownership. The ducks generally separate of their own accord into their own flocks towards evening, and they are then driven home, well fed, and comfortably housed for the night. It is no uncommon sight to see upwards of 3,000 ducklings in one establishment. When the outside accommodation is fully occupied, the cottage or living room is frequently used as a dormitory for ducklings. They are always kept very clean and on dry straw.

Very little description is required as to plumage. It should be of the purest white throughout and fit close to the body. Consequently the Aylesbury duck has advantages over those varieties which have many colours. Wherever colour and markings are considerations, size, stamina, and laying propensities have to some degree been sacrificed for these characteristics, thereby causing diminution in usefulness. Whereas the Aylesbury, being one self-colour, is improved year by year by the introduction of fresh blood. So great is the improvement in this variety that most of the reports in our poultry journals of the Crystal Palace Show of late years assert that the Aylesburys get larger and larger. Therefore I maintain that whilst we are improving this handsome and most useful variety as an exhibition bird, we are also enhancing its commercial value as a table fowl. The bill should be broad and long, coming straight from the skull, and it must be a delicate pale flesh colour. A ferruginous soil will often affect the bill in such a manner that it becomes yellow.

Black marks or spots on the bills are very objectionable, and would disqualify an exhibition bird. The pale flesh-coloured bill of the *pure* breed can be kept right in almost every locality with ordinary attention. Birds for exhibition must be kept out of the hot sun, and should not have too much liberty. I will go more fully into this subject in another chapter. The legs and feet should be bright orange, forming a striking contrast with the white plumage. The body should be long, broad and deep; and the keel long and straight from breast to paunch.

It has been suggested that these keel birds are not the correct type for Aylesburys, but as they have a common origin with the Rouen, I think they are the correct type, for length and depth of keel denotes length and breadth of breast bone, upon which is carried the most valuable flesh. At the same time, I must utter a word of caution against an extravagant ideal for the keel, which should not be represented by an excess of loose bags of skin and feathers trailing on the ground, but should consist of actual length and depth of body. Personally I have been most successful with the keeled birds, and shall continue to breed as many of this stamp as possible. The keel does not show itself so much in the first year as it does in the second and subsequently.

A well-matured drake, well fatted, will weigh 10lbs., but this weight is seldom obtained. I consider seven to nine pounds very good birds, and quite heavy enough for stock purposes.

SCHEDULE FOR JUDGING AYLESBURY DUCKS.

VALUE OF POINTS IN EITHER SEX.

		Points
Head and Eye.	General appearance of head, large, straight and long. Dark and full.	8
Bill.	Long, broad, and straight, forming a straight line, or nearly so, from the top of skull. A drake's head and bill measuring about 6½ inches, duck's about 5¾ inches; colour of bill, pinky white or flesh-colour.	20
Neck.	Long, medium thickness, carried in symmetry with the body.	5
Breast.	Full and deep.	
Keel.	Straight and deep, forming a straight underline from breast to paunch.	10
Size.	As large as possible. A well-matured drake will measure 36 inches from end of bill to end of toes when stretched out flat on a table, and weigh 9 or 10 lbs., and a duck will measure 34 inches and weigh 8 or 9 lbs. Any excess on these lengths and weights should be allowed for as extraordinary merit. ...	20
Plumage.	A spotless white throughout, the drake having two or three handsomely curled feathers in his tail.	10
Symmetry.	Large straight head and bill, well carried on a long curved neck. Long, broad and deep body, with straight underline from stem to stern.	10
Legs and Feet.	Very strong and thick in bone; well set, so as to balance the body in a straight line. Colour, bright orange.	5
Health and Condition.	Eye, bold and clear. Plumage, bright and glossy, having the appearance of white satin. Bill, pinky white, and general appearance very lively.	12
	Total number of Points	**100**

DISQUALIFICATIONS.—Crooked back, wry tail, or any other deformity. Bill any colour other than white or flesh-colour. Plumage other than white. Ducks so heavy behind that in the opinion of the Judge they will not breed.

Rouen Duck, bred by Mr Wm. Bycott. Winner of 1st. Dairy, 1st. Lancaster, 2nd. Palace

Modern Rouen Drake, now the property of Mr. Vincent G. Huntley, Innox Mills, Trowbridge.

MODERN ROUEN DUCKS.

THE original colour and shape of Rouens, male and female, was that of the common Malard and wild duck, but, like all modern domesticated fowls, the fancier has brought them to a wonderfully improved state of beauty, shape, and perfection.

There is no comparison whatever in the Rouens of thirty years ago with our deep-keeled, rich-coloured, clear-billed specimens of to-day. As for the *wild* ducks, they are *tame* specimens indeed in comparison.

SHAPE AND CARRIAGE.

Rouens of both sexes should be very large The drake's head and bill should be as long and straight as possible; that of the duck, long and moderately straight. A dished bill in Rouens is most unsightly. In both male and female the outline of the back should be long and almost straight to end of tail, the curled feathers in the drake's tail of course excepted. The breast very deep, and running with a straight parallel keel, or underline, behind the legs, where the stern gradually rises to a close-carried tail. Old ducks are naturally more square in stern.

COLOUR OF DRAKE.

There are few, if any, of our British fowls that can equal the lustrous beauty, the harmonious and pleasing contrast of the Rouen drake's plumage. The bill should be orange in colour, a very slight shade of green admissible, but in any case uniform throughout, perfectly clear from all black stripes or spots, with the exception of the natural black bean at the tip. During the summer season, Rouens, like other waterfowl, cast the outer skin of their bills, and sometimes this appears in large yellow blotches or scales. At our summer shows this should not distract from the merit of an otherwise perfect drake. The entire head down to the neck ring is a beautiful and most lustrous green.

The white ring encircling three-fourths of the neck should be clear and distinct, clean cut, and neatly defined; a wide ring shows want of quality. The breast colour should be a wide, deep, and dense claret, and as clean cut underneath, from shoulder to shoulder, as the distinctive colouring of a shelldrake. All the undercolour below the wings, from the claret to the vent, should be a clear bluish grey, which is composed of very minutely-pencilled feathers. Drakes do not always moult sound in their claret breasts and clear in the undercolour, and when this is the case they are, of course, unfit for exhibition. In all likelihood, however, they will come out clear and sound the following moult, so they should not be despised.

Down the centre of the drake's back should run a broad dark line, increasing in density and lustre towards the tail, parallel with which, on each side, on the top of the wings and including part of the secondaries, are clear lines of the same colour and shade (or one shade darker admissible) as the undercolour, distinct and clear as possible, and free from cloudy shades or rusty feathers. A darker rim again encircles this below, and includes the lower edge and tips of the secondaries, which blend well with the dark flights. White secondaries are a great failing with Rouens of both sexes, and should be guarded against.

The bow of the wing when spread out, which is covered with short feathers, is cinnamon brown. This colour behind the wing-bars gradually darkens and descends into the flights, which terminates in a very dark brown or black. The only portion of the bow visible when the wing is closed is a narrow strip extending from the shoulder-point to the bars. This in colour is a uniform cinnamon-brown.

The beautiful and characteristic wing-bars, which are the only features in connection with the male and female, are composed of a broad purple-blue band, on each side of which is a narrow bar of black, and again a bar of white. The black and white bars in front of the blue band are a shade wider than those behind, but all must be clear, distinct, a striking and lustrous contrast of colours. The stiff tail feathers of the drake should be dark-brown, like the flights, a very narrow lacing of white on the lower edge of each tail feather admissible. More white than this is objectionable. White or ashy tail feathers denote unsoundness, and white rumps and white secondary wing feathers are sure to follow. The upper tail coverts, the curled feathers, and directly under the tail should be a glossy green-black. The blue-grey undercolour should join on to the black, under the tail, sound and clear. The undercolour here has a decided tendency to terminate too light in colour, frequently an objectionable band of white divides the blue and the black. All breeders have this difficulty to contend with more or less in breeding bright-coloured drakes.

The other extreme is also objectionable in the exhibition specimen. If drakes are too dark underneath, behind the legs, they are generally too cloudy and dark on the back, although these are useful birds to breed from occasionally.

Mr. J. Partington's Rouen Ducks. Winners of two firsts and the challenge cup at the (Crystal Palace, 1891; also first, Dairy, Birmingham, and other important Shows.

COLOUR OF DUCKS.

The down and undercolour of a Rouen duck is black or dark-brown. The upper ground colour almost throughout is also dark-chestnut brown, with a greenish lustre. Each feather has an inner and outer pencilling of rich golden brown. This pencilling should be as even and uniform on the breast, back, underbody, rump, and tail as possible, the exceptions being the head, neck, throat, and bow. The three former are brownish-grey, two light-brown lines running from the base of the bill above the eye. The visible part of the wing bow, as described in the drakes, is a dull-brown, with a single gold minute pencilling round each small feather, resembling honeycomb in shape and regularity. The wing, the wing bars, and the primary flights are exactly the same as those of the drake already described.

A perfect duck's upper bill in colour should be a bright orange with a decided black centre mark, which must neither extend to the base of the bill, the side edges, or within one inch from the end.

Green or lead-coloured bills are very objectionable in both sexes, but during the breeding and moulting season the best marked bills go temporarily dull in colour.

SCHEDULE FOR JUDGING ROUEN DRAKES.

GENERAL CHARACTERISTICS.

		Points
Head and Beak.	General appearance of the head should be massive, and of rich iridescent green, with long, wide, flat beak, well set on, in a straight line from the tip of the eye.	5
Colour of Beak.	A bright greenish yellow, with black bean at the tip.	5
Neck.	Long and tapering, slightly curved but not arched, carried erect in symmetry with body. Colour, rich iridescent green.	5
Ring.	Perfectly white and clean cut, dividing green neck and claret breast, but not quite encircling the neck, leaving a small space at back.	5
Claret Breast.	Rich claret colour, coming well under, clean cut, not running into body colour, and quite free from white pencilling or chain armour.	10
Chain Armour or Flank Pencillings.	Colour, a rich bluish French-grey ground, well pencilled with glossy black, perfectly free from white, rust or iron.	5
Stern.	Same ground as flank, very boldly pencilled close up to vent, finishing in an indistinct curved line (perfectly free from white) followed by rich black feathers up to tail coverts.	5
Tail Coverts.	Colour, black or slaty-black, with brownish tinge, with two or three greenish-black curl feathers in centre.	5
Back and Rump.	Colour, rich greenish-black from between shoulders to rump.	5
	Carried forward	50

SCHEDULE FOR JUDGING ROUEN DRAKES—CONTINUED.

		Points
	Brought forward ...	50
Large Coverts. (Wings)	Pale clear grey. (Small coverts) French-grey very finely pencilled. (Pinion coverts) dark grey or slaty-black.	
Bars. (Wings)	Two, composed of one line of white in centre of small coverts; colour, grey, tipped with black, also forming a line at the base of the flight coverts, which latter feathers should be a slaty-black colour on the upper side of the quill and a rich bright iridescent blue on the lower side, each of these feathers being tipped with white at end of lower side, forming two distinct white bars (the pinion bar being edged with black) with a bold blue ribbon mark between the two.	5
Flights. (Wings)	Slaty-black, with brown tinge free from white	
Colour throughout	Distinct markings, clean cut and well defined in every detail. Plumage bright and glossy.	10
Size.	As large as possible, 36 inches being a fair length from end of bill to end of toes when stretched out on a table, 9 to 10 lbs. being a fair good weight for a matured bird.	10
Symmetry.	Great length, broad and square, very deep in keel, just clear of the ground from stem to stern.	10
Legs and Feet.	Large in bone and well set, so as to balance the body in a straight line. Colour, bright red.	5
Health and Condition.	Bright and glossy plumage, wearing every feather, bold and clear in eye. Heavy in weight but not broken down	10
	Total number of Points	100

DISQUALIFICATIONS.—Crooked back, wry tail, or any other deformity; white flights; no ring on neck; black saddle on beak; leaden beak; wing down or twisted; no wing bars; broken down in stern.

SCHEDULE FOR JUDGING ROUEN DUCKS.

GENERAL CHARACTERISTICS.

		Points
Head and Beak.	As near approaching the drake in size, shape, and form as possible.	5
Colour of Head.	Rich golden, almond, or chestnut brown, with a wide brownish-black line from the base of the beak to neck, and very bold black lines across the head, above and below the eye, filled in with smaller lines between	5
Colour of Beak Saddle.	Bright orange, ground with black saddle extending nearly across to each side of the beak and about two-thirds down towards the tip, with bean as in the drake.	10
Neck.	Long and tapering, slightly curved, and carried in symmetry with the body as in the drake, same colour as head, with brownish wide line at the back of neck from shoulders, shading to black up to head.	5
Ground Colour.	Rich golden, almond, or chestnut-brown, even in colour throughout.	15
Pencilling.	Every feather, excepting wing flights and bars, should be distinctly pencilled from throat and breast to flank and stern, with lustrous black or very dark brown, including back and rump, and wing and tail coverts, with a greenish lustre in black pencilling on rump.	20
WINGS { **Bars.**	Two distinct white bars, with bold blue ribbon mark between, as in the drake.	
Flights.	Brownish, slaty-black, no white.	5
	Carried forward	**65**

SCHEDULE FOR JUDGING ROUEN DUCKS,
CONTINUED.

		Points
	Brought forward	65
Size.	As large as possible, about 34 inches in length, and 8 to 9 lbs. is a fair good weight	10
Symmetry.	Long, broad and square, massive in appearance every way, very deep in keel, square in carriage from stem to stern, but not touching the ground.	10
Legs and Feet.	Colour, dull brown orange. Large in bone and well set, so as to balance the body in a straight line.	5
Health and Condition.	Full plumage, bright and full of lustre. Bold eye. Heavy but not down in stern	10
	Total number of Points	100

DISQUALIFICATIONS.—Crooked back, wry tail, or any other deformity; white ring, or approaching white, on neck; wing down; white flights; no wing bars; leaden beak; broken down in stern; or so heavy that in the opinion of the Judge they will not breed.

MODERN PEKIN DUCKS.

This variety differs from all others in shape and carriage. It has been wonderfully improved since its first importation into this country. It is of Chinese origin, and was first imported into England about eighteen years ago.

The *Monthly Freeman* for March, 1891, says:—"The Pekin duck was brought from China to this country and the United States of America by Captain J. E. Palmer, in the year 1874. The large size, colour, and splendid appearance of these birds caused a great demand for them. Their marvellous egg production, and adaptation for fattening, was the talk of the fanciers, but in a few years they reduced in size and egg production, and also became in many cases unfertile. This was caused by continuous incestuous breeding. Captain Palmer made a second importation. These were brought from Pekin to the coast by Major Ashley, and put on board the vessel. Fresh blood brought back their original vigour, and the Pekins once more regained their original standing and popularity."

The monthly refers to the remote past in the following words:—"The Pekin duck, like the Cochin, is possibly the product of several thousand years' culture. This is shown by the size and form, and almost rudimentary wings. The widest or furthest departure from the wild Mallard At one time it appeared as if it would take the place of all others, but the rage for Pekins did not last very long. Although it looks immense in frame, it is very deceptive in weight, as it is not predisposed to excessive fat. Immense looking specimens will seldom weigh more than seven or eight pounds each."

One authority on the Pekin duck says:—"The flesh, being rather dry, does not compare well with the flavour of the Aylesbury or Rouen's," whilst Mr. Lewis Wright says:—"The flesh is delicate and free from grossness"; and I quite agree with Mr. Wright.

The general characteristics of the Pekin are peculiar to itself, the carriage being most striking of all. It is almost upright in appearance, resembling as near as possible that of a small, wide boat standing on its stern, the bow leaning slightly forward. The head is large and carried well forward. The bill should be short, straight, and thick, and of a bright orange colour. The neck should be thick and long, and when furnished with a nice mane or frill on the top, it adds greatly to its beauty. The body should be as long, broad, and deep as possible.

Opinions differ as to keel, but my own idea is with those in favour of it, for, as with other varieties of ducks, depth of keel denotes depth of breast bone, on which is carried the most valuable flesh; and whether keel is the correct type or otherwise, it will be found useful. I find by careful observation that the birds with this particular point most developed are the birds which have won the principal honours at the majority of our best exhibitions during the last few years. At the same time, excess of keel should be guarded against.

The legs and feet are of a bright orange colour, and are set well back, causing the bird to carry its body in a very erect position, almost like that of the Penguin. This erect carriage is much admired by judges and fanciers, and is sometimes termed style, without which there would not be much chance of winning a prize now-a-days at any of our large shows, for, as a rule, this variety comes up in great force when proper classification is provided for them.

Pekins are non-sitters and good layers. They are also good table birds, putting on flesh and maturing at an early age; but, unlike the Aylesburys or Rouens, they cannot bear confinement. They are very lively and rather wild in their habits, and seem to thrive best when at perfect liberty.

They are difficult to fatten in confinement when pure bred, but when crossed with some other variety the produce grow amazingly, and may be got ready for market quite as early as the pure Aylesbury, and under some circumstances would be more profitable.

They are, undoubtedly, a valuable variety, and have justly gained popularity by their beauty and economic qualities.

Pair of Modern Pekins. The Drake winner of first and cup at the Crystal Palace; first, cup, and challenge cup, Liverpool. Duck winner of first, Crystal Palace; first and special Liverpool; first and special, Gladstonbury, and many other prizes. The property of Mr. Frederick Davis, Woollashill, Pershore.

SCHEDULE FOR JUDGING PEKIN DUCKS.

VALUE OF POINTS IN EITHER SEX.

GENERAL CHARACTERISTICS OF THE PEKIN.

		Points
Head and Eye.	General appearance of head, large; skull, broad and high; cheeks, heavy; throat, slightly gulleted; eye, dark, and partially shaded by heavy eyebrows, and bulky cheeks.	5
Bill.	Short, broad and thick, slightly curved, but not dished, bright orange colour, and free from black marks or spots.	10
Neck.	Long and thick, carried well forward in a graceful arch or curve.	5
Body.	General appearance of the body, that of a small, wide boat standing almost on its stern, and the bow leaning slightly forward. Breast: broad and full, followed in underline by the keel, which should increase in depth between the legs to a broad, deep paunch and stern, carried only just clear of the ground.	10
Tail.	The tail should rise abruptly from the stern, the quilled feathers curving upwards towards the neck, with an immense growth of plumage on the rump, almost covering the quills of the tail. The drake should have two or three handsomely-curled feathers on the top	5
Size.	As large as possible. A well-matured drake will weigh 8 to 9 lbs., and a duck 7 to 8 lbs. Any excess on these weights should be allowed for as being of extraordinary merit. ...	20
	Carried forward	55

SCHEDULE FOR JUDGING PEKIN DUCKS—CONTINUED.

		Points
	Brought forward	55
Legs and Feet.	Legs and feet should be strong and stout in proportion to the size of the bird, set far back, causing very erect carriage, and be of a bright orange colour	5
Plumage.	The plumage must be a sound, uniform buff-canary, or deep cream-colour throughout. It should also be much more abundant in this than in any other variety of ducks. An arched mane on the neck adds greatly to the beauty and character of a Pekin. The thighs and fluff should be well furnished with an abundance of long soft downy feathers.	15
Carriage and Symmetry.	Head: short and thick, carried well forward on a long, thick, handsomely-curved neck. Body: long and deep. Shoulders and back: broad. Breast: very prominent, keel commencing at breast and increasing in depth to paunch, which rises up to a high-carried and well-spread tail. Legs and feet: set well back, compelling the bird to carry its body erect, and in a style peculiar to this variety.	15
Health and Condition.	General appearance: very lively. Eyes: bright and clear. Plumage: bright and glossy, and one even colour from head to tail. Beak, legs, and feet, clean and bright in colour.	10
	Total number of points	100

DISQUALIFICATIONS.—Crooked back, wry tail, or any other deformity; white plumage; black marks or spots on the bill; or so heavy behind that in the opinion of the Judge they cannot breed.

Pair of Cayugas, winners of first prize at the Dairy Show, and first, Birmingham. The property of Mr Fred. Davis.

THE CAYUGA DUCK.

This handsome and useful variety of waterfowl has, during the last few years, gained much favour, as was visible at the late Liverpool show, the classes being well filled with birds of great merit, the really marvellous improvement of this variety being very striking; and there is no doubt that "the large black duck of North America" is gradually becoming more popular. The beauty of their lustrous green-black plumage, the great weight to which they attain at an early age, along with their undoubted economical qualities, their hardihood and prolific laying propensities, are sufficient to recommend them; and not only as a fancy duck, but also as a bird of intrinsic commercial value. It will thrive and keep itself in first-class exhibition form on any soil and in almost any kind of water. It is equally well adapted for exhibition or market purposes, and I have the greatest confidence in recommending the Cayuga both as an ornamental and useful variety of duck.

A large breed of "black ducks" has been known in England since the beginning of the present century, but opinions differ as to their origin; however, it is a matter of perfect indifference to me whether they were in the first place cultivated and made into a sub-variety in England, or whether they were first established on the banks or on the lake Cayuga, from whence they take their title.

There is no doubt they have a common origin with our other large domesticated ducks.

The marked improvement in the Cayugas, both as regards size and colour, is so great, that I have every reason to be well satisfied with them, and there is no denying the fact that they can be kept in perfect health and condition under circumstances which would prove fatal to either Aylesburys or Pekins. In shape they are becoming more like that of a first-class Rouen or Aylesbury. Specimens may now occasionally be seen with long straight heads and beaks, long necks and bodies, and deep breasts, and keels may also be found in some of the best specimens.

Now, with reference to breeding Cayugas, I believe there is still a very great improvement to be made in this variety, and as size is a very important point, and one which will always have my consideration in preference to "green sheen," therefore, I would advise breeders to try one or two experiments, namely, cross your Cayugas with the Rouens, Pekins, or Aylesburys.

If the experiment is made with the former, you may naturally come to the conclusion that the brown colour will take more breeding out, being most closely allied to the original parent stock, the Mallard. If you use the Pekin, the carriage will be the most difficult thing to contend with. If you use a good Aylesbury duck and a Cayuga drake you will find the result very satisfactory, for, as a rule, the ducklings would be either black or white, and would remain so until the second moult, which would of course be in their second year, when the black ones would throw a certain percentage of white feathers, perhaps a few more than they do now, in which case the selection of the next year's breeding stock should be carefully and judiciously made. All the white ones should be killed and marketed as soon as they are fit; also all those which throw too many white feathers during their second moult.

Pure-bred Cayuga ducks often show white feathers on their breasts and other parts of their bodies during the second or third moult. This does not always point to impurity of blood. I have known stock birds showing many white feathers on the breast and on other parts of their bodies whose progeny were all perfectly black during the first year, but during their second moult *they* also showed white feathers.

I do not think the Cayuga can be improved by inbreeding, neither do I think the East Indian would be of any service for this purpose, for they are too small.

The Cayuga duck has undoubtedly been crossed with both the East Indian and the Rouens. The former imparting the most beautiful green-black plumage, but at the sacrifice of size. On the other hand, the Rouens certainly improve the size of the Cayuga, but the affinity for the colour of the Mallard is so great, that it would take years to breed it out again.

I noticed several of the largest specimens at the late Liverpool show with brown pencilling on the breast and throat.

I have no hesitation in saying that white could be bred out in less than half the time it would take to breed out the brown, so great is the tendency to revert to the original parent stock.

The above experiments may appear troublesome to some fanciers, causing, as they undoubtedly do, many sports, and apparently freaks of nature, but in my opinion, such experiments, when carried out in a judicious manner, are the correct way to improve this useful variety of ducks. While if inbreeding or crossing with a more diminutive breed were indulged in for the sake of colour, we should eventually ruin the excellent qualities of the Cayuga. Therefore, let me advise breeders of this variety *not* to cross with the East Indians, and if you try experiments with any other variety, kill off all birds which are not of the stamp or colour you desire.

The ducklings produced by such experiments would be of first-class quality for the table. They would mature early and attain to a good size. The skin and flesh would be white, very fine in texture, and rich in flavour. I have seen Cayugas weighing 15 lbs. the pair at six months old, but they were exceptionally fine birds. They are naturally very quiet in their habits, and seem to do well under almost any circumstances. I would strongly recommend them to gentlemen who have accommodation for a few ducks.

The flesh is quite equal, if not superior in flavour, to any other variety of domesticated waterfowl.

SCHEDULE FOR JUDGING CAYUGA DUCKS.

VALUE OF POINTS IN EITHER SEX.

GENERAL CHARACTERISTICS OF THE CAYUGAS.

		Points
Head and Eye.	General appearance of the head: large, and of a lustrous green colour, with long, wide, flat beak, well set in a straight line from the tip of the eye, which is full and black	5
Colour of Bill.	A slaty-black, with a dense black saddle in the centre, but not touching the sides nor coming within one inch of the end. The bean black.	5
Neck.	Long and tapering, carried in a graceful curve. Colour: a bright lustrous green	5
Body.	Long, broad and deep, breast prominent. The keel coming well forward to breast, which forms a straight underline from stem to stern. Colour throughout, a bright metallic black with as much green lustre as is possible to get. The wings naturally are more lustrous than the rest of the body plumage	10
Tail.	Carried well out and closely folded, and in the drake's two or three curled feathers in the centre	5
Size.	As large as possible. A well matured drake should weigh 7 to 8 lbs. and a duck 6 to 7 lbs. Any excess on these weights should be allowed for as of extraordinary merit ...	20
Legs and Feet.	Large and strong in bone, placed midway in the body, giving the bird a similar carriage to that of the Rouen. Colour: a dull orange-brown	5
	Carried forward	55

SCHEDULE FOR JUDGING CAYUGA DUCKS.
CONTINUED.

		Points
	Brought forward	55
Plumage.	The plumage should be a lustrous green throughout, and as free from purple or white as possible. The whole of the back and upper part of the wings, the breast, and under parts of the body being deep black; brown or purple being objectionable, although not a disqualification	15
Symmetry.	Large head well carried on a long curved neck. Great length of body; broad and square; deep in keel; clear of the ground from breast to paunch. Shoulders broad; back long; tail carried well out; legs and feet well set so as to balance the body in a straight line...	20
Health and Condition.	General appearance very lively. Eyes bright and clear. Plumage bright and glossy throughout, with as much green lustrous bloom on as possible. Beak, legs and feet, clean and bright in colour	10
	Total number of Points	**100**

DISQUALIFICATIONS —Crooked back, wry tail, or any other deformity. Red or white feathers amongst the black plumage. Bill: orange or dished.

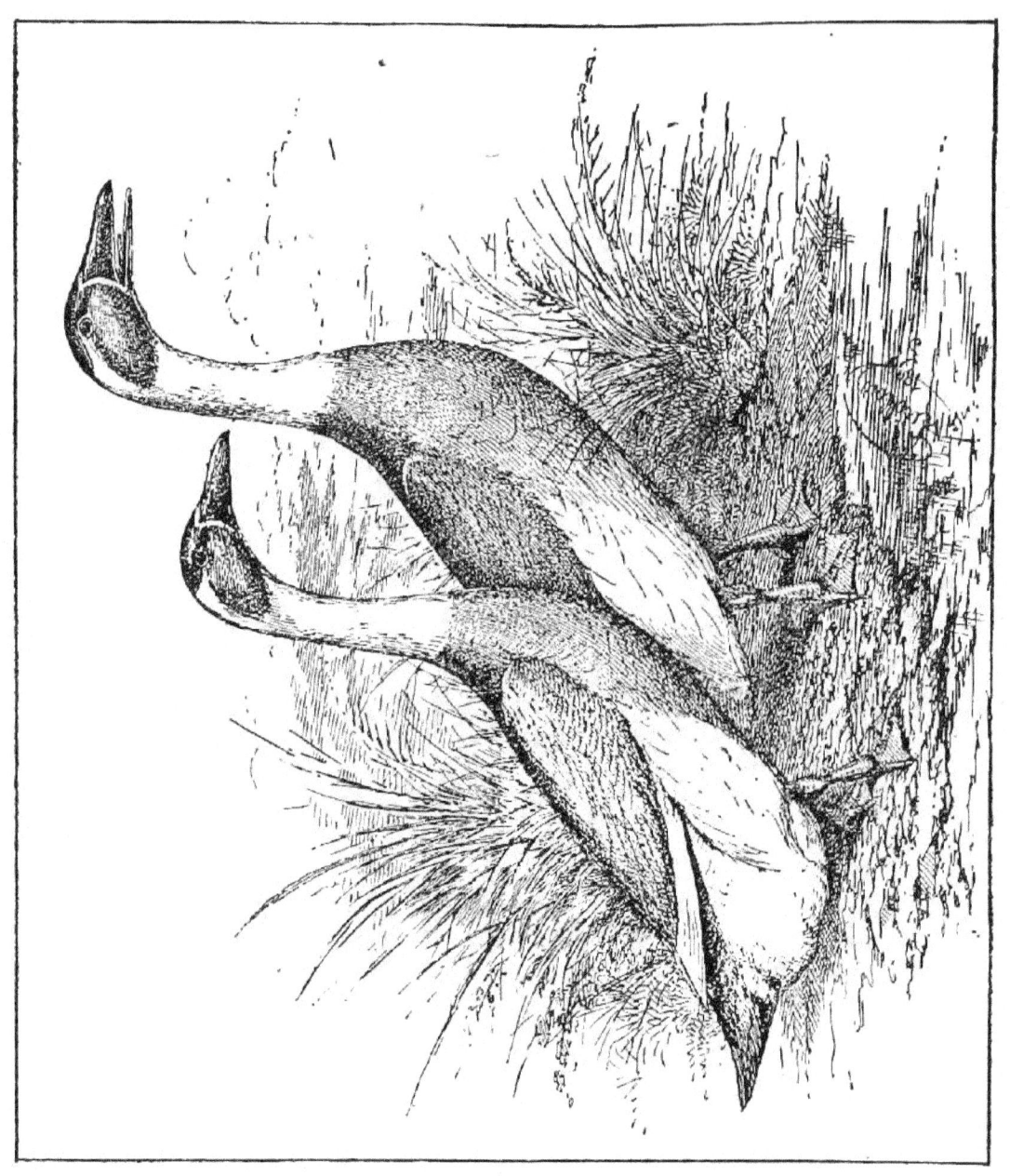

Indian Runner Ducks. Winners of two Firsts and Silver Cup at the Crystal Palace Show, 1907. Bred and exhibited by H. Digby.

THE INDIAN RUNNER.

WITHOUT in the slightest degree wishing to be disrespectful towards or desiring to cast doubt on the apparently irreproachable authenticity or *bona-fides* of any opinions expressed hitherto regarding the introduction into the British Isles of the feathered genus "Indian Runner," I humbly submit that the date, circumstances, and history of the debut of the birds, so far as this country is concerned, is more or less a matter of uncertainty and speculation. Obviously if their title affords any guide as to the land of their nativity,—and it is reasonable at least to rely on the derivation of their distinctive title for the purpose of tracing and fixing their origin,—to India must inevitably belong the distinction of being the birth-place of the original type. Admitting then that the "Indian Runners" are a species, which first saw the light of day in India, the question which suggests itself is "how did they get into this country?"

In an able and comprehensive treatise on the breed, Mr. J. Donald, of Wigton, Cumberland, fixes the date of their introduction into this country as some 50 years back, and explains the circumstances surrounding it in the following sentence:—"A drake and trio of ducks were originally brought from India, by a sea captain to Whitehaven, and presented them to some friends, who at that time followed the occupation of farming in West Cumberland." Mr. Donald also states that "another consignment was imported by the same gentleman some years later, and from these two importations it is probable that all the present day 'Runner' ducks are either directly or indirectly descended."

During recent years Miss Wilson-Wilson, of Kendal, myself, and others have made special efforts to trace the origin of this variety, and, if possible, to import fresh birds of the original stock   I am, however, sorry to say that all efforts in this direction have proved futile. After all, however, as a matter of plain and simple fact, the all important point, and certainly the one with which I and others of the "Fancying persuasion" are more directly and intimately concerned, is not the subject of origin, but the retaining of the pure type and character of an honest conception of the original "Indian Runner." Hence, for the nonce, I shall leave the vexed question of origin and proceed to the consideration of how we breeders can turn the "Indian Runners" to best account both for exhibition and other purposes.

Until very recent time the variety had, for an unbroken period of upwards of fifteen years, been lying in a comparitively dormant state, that is to say, as a "fancy" or exhibition bird. The "Runner" had up to then been obliged to compete in the "Any other variety" classes, at all exhibitions where it entered the lists. No serious organized attempt was made to popularise the species until the end of the year 1895, Before that there had been attempts on the part of individuals to place the "Runner" as a definite species on a sounder and more popular basis, but all these endeavours had not served to give it that attractiveness and popularity, which its more sanguine supporters believed it deserved.

Towards the close of the year mentioned, Miss Wilson-Wilson, at the Dairy Show, sought my opinion as to the advisability or otherwise of making a joint endeavour to raise the "Runner" in the estimation of the "fancy." Needless to say, I advised my "cosy coop" friend to become a member of the Waterfowl Club, in the belief that that influential and well-informed body might be able to assist in drawing up a "Standard of Perfection" for the "Runner" species, and also in procuring the much-desired classification for it at all exhibitions where waterfowl are shown. The credit for the first class of any importance for "Indian Runner" ducks which has been provided at any leading shows for many years belongs entirely to the lady named, for it was mainly through her instrumentality and generosity that twenty-one pairs of "Runners" were drawn together in November, 1896, at Kendal. This latter circumstance has undoubtedly been the chief factor towards introducing several recruits to the Waterfowl Club, and also the means of having caused to be framed an official "Standard of Perfection" for the guidance of fanciers and judges.

That this "Standard" is appreciated is abundantly attested by judges and fanciers alike, the former regarding it as a reliable guide at all times, and the latter finding in it much to aid and instruct them in the "selection," fostering and developing of this particular fancy.

I have deemed it wise to reproduce at the end of my present article a copy of this Standard. As will readily be understood and conceded, the Standard embodies the results of innumerable enquiries and many years' practical experience. The circumstance of my having adopted it as an appendix to this work renders it both unnecessary and superfluous for me now to enter at any length into the characteristics, etc., of the "Runner."

A few words on the chief properties of the birds may not however be out of place. Not only do the "Runners" surpass all other known species of ducks as egg producers, but they are highly esteemed as ornamental waterfowl. Perfect specimens are truly beautiful, but, by no means plentiful or easy to breed. On this, if on no other, they commend themselves to the fancier having a desire to make his hobby interesting as well as remunerative.

To the remarkable capacity of the "Runners" for egg-laying may, in large measure, be attributed to the charm they possess for the small section of duck fanciers who have affected their cultivation hitherto. Experience shows that their season of laying begins earlier and lasts longer than that of other ducks. They stand in this respect in the same relation to waterfowl as Hamburghs have done to poultry.

No claim is made for the species on the ground of their fitness for marketable purposes, inasmuch as the smallness of their bodies and their active habits render them, generally speaking, unprofitable for the table. Although I have observed that "generally speaking" they are not a profitable line to rear and cultivate for the market, I must, however, admit that when young, for those who can afford to breed them for table purposes, they are a choice delicacy, their flesh being finer in texture and richer in flavour than that of most other ducks.

We cannot expect any one breed to possess both unusual laying and table qualities. Breeders of cattle select their stock for the production of milk in one case and of flesh in another, but they do not look for the perfection of both qualities in one animal. The same principle holds good with the breeding of ducks of all sorts. Breeders must choose those strains respectively in which the distinctive feature striven for predominates. In one strain the food is assimilated for the production of eggs, and in another for the putting on of flesh.

The fact of the "Runner" having recently found its way into the yards of fanciers may or may not add to its value as a purely useful duck, but it will most certainly have a decided tendency, and that a wholesome one, to retain for us as well as posterity what is conceived to have been the distinguishing characteristics of the original type of the breed.

"Indian Runner" drakes have from time to time been crossed with ducks of other breeds, evidence of which is frequently found, amongst other things, in the colour of the plumage and the hue and formation of the bills.

The "Runners" are, as already conveyed, most active in their habits, capital foragers, and on a good run are able to find three-fourths of their own sustenance.

It is only on rare occasions they evince a desire to sit, and when they do so it is not wise to trust them with eggs, for they cannot be relied upon for successful incubation.

When young they are in no sense tender birds to rear, in fact, the ordinary attention bestowed on ducklings of other breeds suffices to bring them to a sound maturity.

Reference to my illustration and the "Standard of Perfection" which follows will, I hope, enable my readers to make proper selection either for breeding or exhibition purposes.

Colour and markings being almost identically the same,—in both drakes and ducks,—renders it unnecessary to keep two pens of stock birds, this fact being a great advantage to the fancier whose space is limited.

STANDARD FOR INDIAN RUNNER DUCKS.

(As approved by the Waterfowl Club.)

VALUE OF POINTS IN EITHER SEX.

GENERAL CHARACTERISTICS.

		Points
Head, Shape of Eyes and Bill.	Fine and comparatively flat, with the eyes situate high up. Bill strong at the base, broad and long, coming as near as possible straight down from the skull, giving it the appearance of a long wedge. The colour of the bill when young is yellow, but as the birds grow a green line begins to develop at the base of the beak, and this is frequently accompanied by green spots, which gradually increase in number and size, until by the time the birds are a year old the whole surface of the mandible is entirely green; a black bean on the tip is preferable.	15
Head Markings.	The head should be adorned with a cap and cheek markings, as near the same colour of the body as possible, a narrow white line should divide the cap from the cheek marks, and a line of white about $\frac{1}{8}$ of an inch should divide the base of the bill from the head markings.	10
Neck.	As long and thin as possible, perfectly white from the head to where the breast marking begins, which should be about $1\frac{1}{2}$ to 2 inches from the base of the neck.	10
Body.	Long, narrow, and racy looking, without the slightest indication of keel	10
Legs.	The legs are a deep bright yellow colour, set well back, compelling the bird to carry its body erect something like the form of a penguin.	5
	Carried forward	50

STANDARD FOR INDIAN RUNNER DUCKS—
CONTINUED.

		Points
	Brought forward	50
Body Markings.	Whatever colour an Indian Runner may be, that colour should be uniform throughout the whole of its surface plumage, except the tail of the drake, which is darker. The breast should be evenly cut about half way between the point of the breast bone and the legs. The shoulders, top part of wings, and tail should be of the same colour as the breast, and should be the shape of a heart pressed flat on the back. Flights and fluff, white, except an indistinct line of colour which runs from the base of the tail to the thighs.. ..	25
Symmetry.	General appearance, carriage and condition.	25
	Total number of points	100

Colours preferred. Fawns and Greys. Weights not to exceed 5 lbs. in either sex; a fair average weight for drakes is $4\frac{1}{2}$ to $4\frac{3}{4}$ lbs., and for ducks 4 lbs. 30 inches is considered a fair good length for a drake not exceeding 5 lbs., whilst 25 inches is considered a good length for a duck not exceeding 4 lbs; any excess on these lengths in birds not exceeding the above weights should be allowed for as extraordinary merit.

It is the nature of the **true** Indian Runner duck to **run** without waddling, like most breeds of ducks, but unfortunately, this distinguishing characteristic cannot be seen in the Show pen, for it is only observable when the ducks are at liberty.

DISQUALIFICATIONS.—Blue ribbon wing bars, claret breasts, horizontal shape or carriage, absence of feathers from the flights, secondary flights, or any other part of the body; slipped wings, wry tail, or any other deformity.

BREEDING AND TREATMENT OF BREEDING STOCK.

If due regard has been paid to the "selection of stock," having paid sufficient attention to the variety to be kept, and also the object to be attained, breeding will be found a very important part of the business; in fact, the part which will bring the "grist to the mill," if carried out in a proper manner.

In order to make ducks pay a good return it is absolutely necessary to fix upon the object to be attained. There is not a shadow of a doubt that breeding exhibition ducks and geese has been very profitable to many farmers (whose names I could give.) Neither is there any doubt that as they are reared in Bedfordshire and Buckinghamshire for the London market, that they can be and are annually produced most profitably.

If the breeding and rearing of ducks and geese were better understood and developed into a systematic industry, there is not the slightest doubt in my mind that their culture could be made the means by which substantial profits could be made. Ducklings for market should be bred as early as possible. They should be raised under the most favourable conditions to ensure rapid growth and early maturity, and they should be killed when the food that has been given to them has produced the best results. Before they begin to cast their first feathers they should be marketed or sold to private customers at the season when the demand is greatest, and consequently the prices highest.

In order to be able to get eggs to produce early ducklings, it is necessary that your breeding stock should be young, healthy, vigorous, medium-sized birds. They should always have perfect liberty, and should not be exhibited at all. If you disturb your breeding stock by sending them to shows, you will certainly have to wait weeks longer for your eggs for sitting. Therefore I would advise all breeders to let their stock birds stay at home. Do not on any account send your breeding ducks unaccompanied to shows. If you will allow them to remain at home their progeny will not only be stronger but more numerous. If you send them long distances and friendless to exhibitions your chances of winning are only very small, for when they arrive at the show they are at the tender mercies of the committee, who will very often forbid water or food being supplied until after judging.

I have often complained aloud in reference to this matter, and in one or two cases I have been the means of food and water being supplied before judging.

When waterfowl are sent to shows, and subjected to the usual treatment, their constitutions are to some extent impaired, and many really first-class stock birds have been reduced to absolute sterility by being over-shown. They have been returned home with empty honours, their whole system broken down by a desire for fame which was almost impossible for your otherwise first-class breeders to obtain, and they have been ruined and made worthless for the purpose for which you bought them.

In these days of keen competition the exhibition bird is *one* thing, but the best stock birds, for producing these *monsters* quite another. If you wish to breed ducks and geese for exhibition, and you have a desire to surpass all others in the show pen, do not be in a hurry to hatch very early birds.

I find that ducks and geese hatched in April or May grow very fast and make plenty of bone in a very short time, and these are the birds which eventually make the monsters by the time of the Crystal Palace Show. They have all the warm weather to grow in, and their nature appears to take advantage of it.

Personally, I have always been most successful with birds hatched in May. If your stock birds are laying a satisfactory number of well-formed and properly shaped eggs, let well alone. but if they are only laying indifferently or if the eggs are irregular in shape, appearing rough or thin in shell, or, as is sometimes the case, without shell altogether, or should they lay double-yolked eggs, or produce two eggs in one day, or be found guilty of other irregularities, then you may be sure there is a reason for such irregularities, and try to discover it.

Now, as a rule, most, if not the whole, of these freaks point to one common fault, and that is feeding your breeding stock too liberally on stimulating or improper food. Should you be troubled by any of the above annoyances just take the trouble to see if you cannot find out the reason why. If you succeed in finding out the reason for the defects, remove the cause, and the disease will disappear.

In nine cases out of ten you will find that your breeding stock have been living too well, in which case all stimulating foods should be discontinued, and only the very plainest kinds of food given to them, especially when your breeders have a good run on grass land. It is very important that your stock birds should be supplied with an abundance of shell-making material, such as lime, chalk, shells of different sorts, ashes, old mortar, rubbish and gravel. All these should be supplied to waterfowl regularly, but more especially during the breeding season.

Having procured a really good breeding pen of ducks from one or two reliable breeders, and which are about or a little over the average size, put them on the run on which you intend they should remain until you have got all the eggs they will lay. Waterfowl should not be removed from one run to another during the breeding season, for when they are disturbed they frequently cease laying for a time. Therefore, in order to get the best results attainable, it is requisite to give your stock birds undisturbed peace at home.

Individual ducks of one family vary considerably in their prolificness, and I have owned birds which have not laid more than thirty eggs in a season; whilst others of the self-same family have laid 150 in a season. I find the average number of eggs laid by birds of one and two years old is about 60 per duck.

Darwin says:—"The wild duck lays from five to ten eggs; the tame one, in the course of the year, from 80 to 100." But if ducks have not been harassed by being exhibited, the average number of eggs will be much higher.

Stock ducks should always be shut up in their house at night, but they should have liberty until it is almost dark, for it is in the evening that worms, etc., come to the surface of the ground, and are at once consumed by the ducks. I am sure I need not tell anyone acquainted with the habits of ducks how greedy they are when on their evening rambles in search of animal food. I have seen my stock birds come home at dusk scarcely able to waddle and refuse the very best of corn if offered to them.

Stock ducks on good grass runs require very little hand feeding, but as it is necessary to keep them shut up until about eight o'clock in the morning, they should be supplied with a fair good breakfast of soft food. They may be tried with a handful of oats, wheat, or barley in the evening. In many cases this will not be required—that is when they have an unlimited grass run.

There are several reasons why stock ducks should be housed at night; and, first, it is protection from foxes and other enemies of the poultry yard; secondly, because they glory in stealing away and selecting their nests in any out-of-the-way or secluded corner; and when allowed perfect liberty, as many farmers' ducks are, they will often drop their eggs into the water, when they sink, and nothing is known of their presence until perhaps some weeks afterwards, when having become rotten they rise to the surface. Thus scores of duck eggs are lost yearly by careless farmers, who think it is too much trouble to shut their ducks up for the night.

It should also be borne in mind that stock ducks should be provided with a good bed of clean, dry straw, for they often drop their eggs at random all over the place, very often ignoring all nest accommodation which may have been provided for them. The bedding should be removed and shaken out daily, and the house swept out. A good swill with water now and again will do the house good.

The best material for bedding is decidedly straw, and when it can be procured at a reasonable price I would certainly advise its use. I prefer wheat straw, but barley, rye, or oat straw will answer the purpose.

When straw cannot be obtained, some other kind of sweet dry litter should be used, such as bean or pea straw, coarse hay, bracken, rushes, or dry leaves, anything which would necessitate a regular cleaning out would be preferred to peat-moss litter. I have tried it for waterfowl, and my opinion of its use is decidedly against it, for in consequence of its being absorbent, it looks cleaner than it really is. It has also a tendency to encourage laziness, and I am sure no poultry man or boy would like to be accused of such a fault. Therefore, avoid peat-moss litter for ducks.

If you wish your waterfowl to keep in good health and condition, sawdust or wood-turnings, or, as they are sometimes called " chippings," are very cheap, and much preferable to peat-moss, for you can see when they are dirty.

Filthiness in houses is the chief cause of many contagious and deadly diseases, and the use of all material calculated to harbour filth should be carefully avoided. It does not follow that fresh bedding is required every day, simply because I advise its daily removal. If the weather is fine throw the straw into the open air to dry. After a while give it a good shake, and it is then in a fit condition to be returned to the house for bedding.

If the manure and the soiled bedding from waterfowl is made into a tidy little stack, and allowed to eat and rot for a month or two, it will make capital manure, and will grow kitchen garden produce to an enormous size.

Stock ducks should always have access to water. It is their natural element, and they look better when they are allowed free access to it, not only this, but they lay better and their eggs are more fertile.

I would not advise anyone to keep waterfowl unless they can let them have a sufficient supply of fresh water for a bath daily. A constant stream or a good pond being preferable to most artificial arrangements.

INCUBATION.

This branch of the business requires the greatest care and attention on the part of the owner or breeder of waterfowl.

If you wish to obtain the best prices for your ducklings, you must put your first eggs down for incubation as soon as ever your ducks begin to lay. I prefer large Brahmas, Cochins, Dorkings, or cross-bred hens for this purpose, to either ducks or incubators. During the winter and very early spring I usually sit two hens on the same day, for it frequently happens that the eggs are not fertile when produced in very cold weather.

I never give more than nine duck eggs to one hen during the first three months of the year; this number being quite as many as any ordinary hen can comfortably cover and keep thoroughly warm. I usually test the eggs when they have been under the hen seven or eight days, and if I find many clear. *i.e.*, unfertile, I put all the fertile ones under one hen, and procure another sitting hen, then re-set the one which has been deprived of her eggs in consequence of the examination; by this means I am able to continue sitting two hens weekly, by the purchase or hire of one additional hen. I continue this process so long as there is very severe weather, and I find the system practically useful.

As the weather becomes warmer, the quantity of eggs laid will be considerably increased, and with them fertility greatly augmented. In consequence, sufficient sitting hens may be difficult to procure. If so, an "Incubator" may be found not only useful but necessary. Although I frankly admit that I prefer the use of hens to any incubator. The "Westmeria," which I am now using, is so easy to work, and does that work in such an efficient manner, that I have every confidence in recommending it.

The moisture arrangements are, to my mind, perfect. This, of course, forms a very important part in hatching duck and goose eggs.

Again, let the eggs be fifty or one hundred in number, they can be turned within a minute, and when the process of turning has to be done daily this is a great consideration. The temperature is kept at a certain point automatically, and scarcely varies one degree. Some people may, perhaps, look at the cost (others are obliged to do so) of an incubator, but then it is not, like the clucking hen, to be purchased afresh every year.

A fifty-egg incubator should form part of the stock-in-trade of every breeder of waterfowl. They have certain advantages over hens, for they will hatch thin-shelled eggs, which the hens would certainly break. Should a sitting hen fall sick or die on her nest, an incubator is always ready in case of such accidents. They are also very valuable at the time of hatching. It sometimes happens that a hen will crush an egg or two a little before the time is up, and if there were no incubator at work, the gosling or duckling would inevitably be lost, whereas, if you have an incubator at work it is just the thing required, and is able to hatch out those birds which would otherwise be lost.

The two "Westmerias" which I have now working have paid their cost in nothing else than hatching birds out of damaged shells. These machines are so well put together, and so perfectly packed to withstand the effects of the heat, that I think I should be quite safe in saying that they will last me a considerable time.

An incubator is most useful for testing your first batches of eggs, in order to prove their fertility, before wasting the time of hens, or before selling eggs to customers. It sometimes happens that some individual ganders and drakes are absolutely sterile, and for that reason breeders ought not to sell eggs from any particular pen of stock ducks or geese until such breeders have proved the fertility of their stock birds.

INCUBATION.

When duck or goose eggs have been in an incubator or under a hen for seven or eight days it is quite easy to judge whether they are fertile or not. The simplest and best way of testing the eggs is to take them out of the machine, or from under the hen, light a gas jet or a candle in a dark room, take the egg in your right hand, and with your two forefingers and thumb hold the egg before the light, taking hold of it by the small end. Then place your left hand across the top of the thick end of the egg, shading the light from your eyes. If the egg is fertile you will be able to see a clear space at the thick end of the egg, and the lower part will be quite dark. Should the egg appear quite clear, and almost transparent, you may depend it is unfertile, and that it would be only a waste of time to return it to the hen or incubator.

Nests intended for sitting hens with duck or goose eggs, should, as far as possible, be made on the ground. The earth should be hollowed out, not too deep, still it should be made roomy enough to admit a good wisp of soft hay. Hay is much preferable to straw for this purpose, it being much warmer. My reason for advising the nest to be made on the ground is, that duck and goose eggs require more moisture during the time of incubation than other kind of poultry eggs do.

If it is not convenient to make the nest on the ground, a box may be used for the purpose, the bottom of which should be well covered with grass sods or loose damp earth. A large pailful is not too much for the purpose. Two pailfuls would be preferable, taking care that all the corners are well filled up with earth, and then made comfortable with hay.

Eggs when under the process of incubation should not be sprinkled with water. Scores, yes hundreds, of birds have been killed in the shells by this thoughtless practice. Nevertheless, moisture must be supplied to the nest or its surroundings, for, in a state of nature, the sitting duck or goose would leave her nest early in the morning, when her plumage would become wetted by the rain or by the dew which would be on the grass, or, perchance, she may have taken a bath in the pond or stream. In any case her plumage would be wet on her return to the nest, and, further, the nest would doubtless be made on the ground in some nicely shaded situation. Therefore it will be seen that the eggs of waterfowl in a state of nature are damped daily by the natural incubator.

I find the best method of supplying moisture to eggs when under hens, etc., is to take a garden wateringcan with a rose on, and water all the surroundings of the nest, and even the nest itself, with water at about 100 degrees. This should be done at night, when the hens are quietest. Care should be taken not to disturb them, or you may find a difficulty in getting them back to the eggs. By this method moisture is supplied in proper quantities, and at a proper temperature. The shell and inner membrane gradually becomes more brittle, and the young ones are able to extricate themselves from their prisons without risk of losing their lives through inability to break through their prison walls.

If there is sufficient moisture in the ground on which the nest is made, the heat of the hen's body will draw moisture sufficient for all requirements. If the nest is made as advised, the eggs may be carefully taken out of the nest whilst the hen is off feeding, and the nest well watered as above, after which the eggs may be returned to the nest. This method should be repeated three or four times during the last fortnight of incubation.

Moisture is one of the most essential requisites for the successful hatching of waterfowl. Hens should always be sat at night. They seem to take to the eggs more readily and quietly than when sat during the daytime.

It may appear strange to some of my readers that I should write on the "Hen" for incubation of ducks and geese. As a rule, ducks and geese are not very good sitters, and worse mothers, especially high-class birds.

My Aylesburys seldom evince a desire to sit, and if they did, I could not afford to allow them to do so, for I require all the eggs I can possibly get from them.

Toulouse geese are only very indifferent sitters and mothers, so much so that I never risk valuable eggs under them.

There is no doubt that habit has something to do with this partially lost quality, and if we fanciers wish to develop any particular habit or point, we must encourage our pets to develop the same; but if, on the other hand, we do not wish our ducks and geese to indulge in their natural propensity of incubation, and we annually deprive them of their maternal habit, we may come to the conclusion that in course of time our birds will not only become indifferent sitters and mothers, but that they will eventually lose all desire for incubation, for the force of habit is transmitted from one generation to another. I have not the least doubt that some strains of ducks will at no distant period lose all desire to sit. Do not be afraid of sitting too many nests of duck eggs, for if you sit all the eggs your three or even six ducks will lay, there will be no danger of getting too many really first-class ducks.

A "Topper" only comes occasionally. Therefore set all your eggs, and you will find that there is no necessity for a separate breeding pen in order to produce a large number of ducklings for the market. If you set all the eggs your good stock birds will produce, and the season is at all favourable, you may naturally expect a good crop of ducklings, which will, of course, require some care and judgment bestowing on them; but I will go fully into the care and treatment of ducklings both for market and exhibition in the next chapter.

The time required for incubation is twenty-eight days for ducks and thirty days for geese, but different strains or families vary considerably in this respect. I have frequently known ducklings hatch on the twenty-sixth day, whilst others have not appeared until the thirtieth day, and I have also noticed similar variations in the time of hatching goose eggs, some goslings appearing on the twenty-eighth day of incubation, others taking as much as four days longer. Consequently, it is well to be very careful during the time the young ones are hatching. As a rule, it will be far better to leave the sitting hen alone until all are hatched, further than to remove the shells or any other objectionable matter there may be in the nests.

Some breeders remove the goslings or ducklings from the nest as soon as they are hatched, and put them in old hats, small baskets, etc., and then place them on the hearth or kitchen fender, and when the whole are hatched place them under the hen. Such practices and other unnecessary interferences between the attendant and the sitting hen are the causes of the death of hundreds of goslings and ducklings annually.

It is very much better (at all risks) to leave the newly-hatched birds under the hen, undisturbed, for at least twenty-four hours after they are hatched. Nature has provided for their sustenance for fully this time after they emerge from the shell. Therefore, the less interference between the attendant and the sitting hen the better will it be for all concerned.

BREEDING AND REARING DUCKLINGS FOR EXHIBITION.

Ducklings for exhibition should be treated in a very different manner to those for market purposes, and the treatment must be varied according to the number kept.

Experience teaches me that it is much easier to rear a few to a high standard of perfection than it would be to rear a large number. Therefore, I maintain that the person who keeps *only* one variety, concentrating his capital and energy on that one variety, is better able to discover excellencies or faults sooner than the person whose energies and capital are divided into many different sections.

When my ducklings are hatching, I always make it my business to attend to them personally. I do not remove the little creatures from their warm, cosy nest. I simply remove the shells very carefully, without exciting the hen if possible, and should any of the shells happen to be fractured in improper places, I remove all such and give them a bath in warm water, and as I always keep a vacancy or two in one of my "Incubators," I place any such fractured eggs in it, thus often saving the lives of those which sometimes turn out to be my most valuable ducklings. If I had no incubator, I should give these fractured eggs a bath just the same, and then wrap them up in a flannel which had been dipped in hot water, then place the egg in the most convenient place near the kitchen fire, keep it reasonably warm, occasionally renewing the hot flannel, and when once the duckling has quite cleared itself of the shell, I should substitute a warm, dry flannel until the duckling had regained its strength. I should then return it to the hen, for I am fully persuaded that the nest is the proper place, and that the hen or duck whose task it has been to sit on the eggs for thirty days is the proper nurse for ducklings.

I do not believe in taking them from the hen, for I never found either ducks or geese grow to an extraordinary size which had been taken away from the hen at an early age.

Some writers tell us that ducklings do not require brooding. This was astounding information to me, and although it may be true that ducklings do not take quite so much shelter under the hen as chickens do, it is also true that they *do* require brooding. If the hen is taken away from them at an early age they never grow very large, and if a duckling once makes a stand, should droop, or refuse its food for a few days, you may mark that bird *off* as being of no use for exhibition, and as soon as ever it is ready for the market, let it go.

Ducklings reared with a view to exhibition must grow and do well until they are six months old, otherwise they cannot compete successfully against those bred and reared by experienced breeders.

Theory is all very well, but the kind of theory which I have seen advocated recently, concerning the rearing of ducklings, would, in my opinion, prove most disastrous, especially for exhibition.

It is a well-known fact that all kinds of domestic ducks have improved during the last few years, and I would like to ask, whether this improvement has been brought about by assisting nature or by going directly in opposition to nature's rules and laws? If this improvement has been brought about by those principles which are in opposition to nature, how is it that these writers and breeders have not been able to compete successfully with other breeders and exhibitors who believe in attending nature's laws?

I am strongly in favour of leaving ducklings with the hen or duck that has hatched them. I find they are stronger and grow much faster than those which have been removed. I would prefer half the number and have them well nursed in this manner.

I never remove my ducklings from the nest for at least twelve hours after hatching, and if the whole lot hatched evenly, that is, about the same time, I should let them remain in the nest for twenty-four hours. I should make the nest as flat as possible around the edges, so that if a duckling should come out it would be able to get back to the hen without the risk of being starved to death. Of course, I am aware that there are some hens which are very excitable, and these creatures sometimes trample on their charge and kill them; but such accidents as these are often caused by the attendant, who, when he goes to the nest to examine the eggs or the ducklings, receives a sharp peck from the hen in defence of her brood. The attendant does not use sufficient discretion, but frequently gets cross with the hen, and sometimes uses her very roughly, increasing the poor hen's excitement to such an extent, that when she is put on to her nest again, she tramples upon and kills perhaps half of her brood.

If you wish to be successful in this particular part of rearing ducklings you must be very patient, and on no account go to look at a sitting hen after you have had a family quarrel!

I will now suppose that your ducklings are all hatched, thoroughly dry and well brooded, and that it is time they were removed from the nest in which they were hatched, for fear of vermin of various kinds might possibly have taken up there quarters close by.

Remove hen and her ducklings to the place which has, of course, been prepared for them; and first of all, feed the hen with her usual food to her entire satisfaction.—After feeding the hen, it will be quite soon enough to feed your ducklings. I like a large smooth board to feed on best of all; I *have* used an old sack, or a piece of canvas for this purpose. It is a matter of perfect indifference what it is, so long as something *is used* to prevent waste of good food. Throw your egg and bread-crumbs on to the board or sack, and your ducklings will soon learn the art of converting one kind of food into another. Continue the hard-boiled egg, chopped and mixed with twice the bulk of stale bread-crumbs. Feed about every two hours for the first week. Let them have a shallow dish filled with water and a bit of sharp grit put at the bottom. Green food of some kind should be supplied regularly. A good grass sod, 15 or 18 inches square, will be very much appreciated by your ducklings, and the soil will assist digestion for the first week or two, after which sharp grit *must* be supplied.

Almost any kind of greens grown in a garden will answer the purpose. Lettuce is my favourite vegetable for ducklings. I grow large quantities every year, sowing about every three weeks. *Malta* and *Coss* make the largest plants. The youngsters are very fond of it, and I have great faith in it keeping them in health and condition. It gives a tone to the system, and acts as a tonic. Savoys, cabbages, onions, radish-tops, and even weeds will be devoured if your ducklings are kept in confinement. All kinds of vegetables should be chopped up fine with a knife or in a mincing bowl, and supplied once or twice daily, and should be continued until you are prepared to give your birds perfect liberty. When your ducklings are a week old, you may safely dispense with the egg and bread, and use some other cheaper food, which should be a cooked preparation. I know of no better food than "Spratts Patent Poultry Meal." The birds eat it with avidity, and it seems to grow plenty of bone, which is very essential for exhibition ducks. The way I prepare this food is a little more trouble than some people would care to take. Still I believe in the proverbial saying, "That if a thing is worth doing at all, it is worth doing well."

Having procured a bit of coarse flesh meat of some kind or other, or a few handfuls of "Prairie Meat Crissel," I boil it in a saucepan or copper pan just as occasion may require, and when thoroughly boiled, pour broth and meat into the poultry meal, and only have sufficient meal to leave the mess very moist. When thoroughly scalded, add barley meal and thirds, or ground wheat, and do not forget a bit of salt and a little bone meal, sufficient to mix the mess into a nice stiff, crumbly paste.

The food now being prepared is ready for service to your ducklings, and as I am dealing with birds intended for exhibition, I will endeavour to show my reason for feeding on boards, sacks, or on the grass, in preference to feeding in troughs. When ducklings are attended to by a careful and intelligent attendant, whether he be the owner or a servant, he will take the food in a vessel of some kind, say a pail, and when he comes to a flock of ducklings, no matter how large or small, he will not throw a large quantity of food down at once, but he will take a handful at a time and throw it broadcast. If the food is properly mixed there will not be one particle of waste, for the youngsters will run about in search of the least crumb. The feeder will of course continue his work by throwing handful after handful until such time as he thinks the birds have had sufficient. It occurs to me now, that whilst you are feeding in this way your ducklings are all the while on the move, and during the process of feeding you are able to see if any of them are ailing, and if so, attend to them at once. Your eye has become acquainted with every individual bird, and long before the wasters are ready for killing your mind is made up which are the wasters and which are the best birds for exhibition. So that when the time comes to draw out the birds for market or killing, you have no difficulty in doing so, for your mind is already made up, and made up in such a manner that a whole day's examination would not be equal to it, and you are able to make your final selection without fear that you are killing a really good bird, and by being able to do so, you are saving time and money. A little extra time spent in feeding will very often decide the question of loss or profit.

Feed with food as above, about five or six times daily until they are a month old. The sooner you can give them their first meal and the later you give them their last is all the better, for when we remember that the food with which they are fed is partially cooked, and consequently made easier of digestion, also, that the small quantity of food a young duck is capable of consuming at one time, it must therefore follow that the contents of the crop must be exhausted very early in the morning. As our object is to grow them not only as fast as possible, but to get them to an extraordinary size, we must supply "fuel" in order to keep up the steam, therefore I lay great stress on a feed *early* in the morning and *late* at night.

After feeding with soft food at dusk, shut your ducklings up for the night. Leave a large shallow pan, either made of tin, iron, or earthenware, fill it with water, and add a bit of good wheat, say a small handful to each duck, and a bit of gravel, that is after they are three weeks or a month old. If they do not clear it out the first night or two, they will soon find out that the wheat is intended for their use, and show you how much they appreciate your kindness by clearing out every grain long before morning. As your ducklings get older you will find that their appetites increase, and they will be able to eat considerably more at a meal, and as a matter of course you will give it to them. You will also find that in consequence of their increased capacity for food they do not require feeding so often. Three times a day will be sufficient after they are six weeks old. I do not believe in any kind of spices or over-stimulating foods for ducklings, except in very damp, cold weather, when I have used a little Poultry Spice to great advantage. It is a preventive of cramp and other diseases caused by damp ground, etc., after rain. By the time your ducklings are ten weeks old you will no doubt have selected and kept a few of the very best, and of course put the wasters in your pocket, if not, why not? for it is quite time all this was done. The selected ones will now require special attention, for they will soon begin to cast their first feathers, and should be treated as per my articles on ducks when moulting.

Good sound food twice daily will now be sufficient for them. Spratts Patent Poultry Meal scalded with the broth, in which some kind of flesh meat has been boiled, or some good "greaves" or a little "Prairie Meat Crissel" mixed with ground wheat, barley meal, or thirds, will make an excellent food for the morning, and a bit of good English wheat for the evening. Not forgetting the green food and the grit, also a comfortable house, well bedded with clean straw, to sleep in.

I have said nothing about water for ducklings for exhibition, for only just lately I have read an article by a writer who distinctly tells us that ducklings will do as well without water for ten weeks as they will with it. This assertion appears to me most unreasonable and unnatural. I have bred a few ducks in my time, perhaps not the thousands that the writer referred to would have us believe he has bred. Still I have bred sufficient to be able to say, fearless of contradiction, that ducks do not thrive or do better without water than with it. I cannot get over the fact that water is the natural element for ducks, and if it is natural for ducks to swim, why should it be unnatural for ducklings? I will not dwell further on this point, for I feel sure that all lovers of water-fowls will let their birds have access to it wherever practical, and if they do not, the result will be cramp, liver disease, death from sunstroke, etc. If you wish to be successful in breeding and rearing ducklings for exhibition, try to assist nature by the exercise of your own knowledge and learning, and do not be misled by anyone who would advise you to work in direct opposition to nature's teachings.

REARING AND FEEDING DUCKLINGS FOR MARKET.

This branch of the business has not had my special attention, and for that reason I do not profess to approach perfection.

I am perfectly well aware that it would not pay to treat ducklings for market in the manner I described in the last chapter for exhibition purposes, certainly not. If your sole object in rearing ducklings is for the market, there will be no necessity for giving high prices for stock birds. Breeders of exhibition birds have frequently a few rough or faulty birds, which they are generally open to sell at a nominal price, and which would answer this purpose admirably.

If my only object in breeding ducks were for consumption, I should never think of breeding from very large famed parents, but rather than this, I should select my breeding stock from a good prolific strain, weighing about or a little over five pounds per bird when in store condition. From such birds it is easy enough to get ducklings to weigh four or five pounds each at ten weeks old, and in some districts these will realise quite as much money as those weighing six or seven pounds each. I have often been told when asking ten shillings per couple for ducklings of extraordinary size, that the price was very high, for they were only ducks after all. Therefore, in order to breed and rear ducklings solely for the market and to a profit, discretion must be used in the selection of breeding stock as well as in the rearing of the ducklings. If your breeding stock are of a prolific strain, and have not been forced to an extraordinary size, there will not be much fear of sterile eggs, especially if they have a fair good run, and the progeny as a rule are very hardy, having sound constitutions to enable them to withstand the strain of forcing and feeding, and being kept from the water for nine or ten weeks, at which time they are or ought to be quite fat and in the very pink of condition for killing, and should be disposed of forthwith.

My treatment for ducklings for market would be just the same as for exhibition for the first five or six days, but after that it would be entirely different, for the growth of immense bone and a sound constitution are of little or no consequence. All we require is a nice plump duckling fit to kill at nine or ten weeks old, sooner if possible, and it only remains for me to show how to rear and feed a duckling up to this said point with as little expense as possible. If I had a flock of ducklings destined for killing, I should not use any kind of fancy foods. All they require is judicious feeding. Care must be taken to keep them clean, warm, quiet, and comfortable, and as regularity is the sole of business, so it is in the preparation of ducklings for the market. I should feed on bran scalded with the broth in which some very course cheap offal had been boiled, giving offal and all. Not that there is anything very feeding or fattening in either of the above, but the bran keeps the bowels in order for the time required, and the offal tempts the appetite and assists the digestion of the ground oats, ground wheat, barley, buckwheat, or Indian meal, which are all good for feeding ducklings and preparing them for an early market. Any of these meals may be used with advantage, mixed with the scalded bran into a nice stiff crumbly paste, and given to your ducklings as often as appetite may require. The times for feeding will be about the same as laid down for exhibition. Whole grain should not be used for this purpose. A plentiful supply of grit should always be within their reach, and a reasonable allowance of water to drink both before and after feeding. It seems to me a cruel thing to deprive a duck of a drink of water until it has gorged itself with food. I think digestion would be assisted if a drink of water were allowed both before and after a good meal As for a swim, well, I am fully convinced that a good wash and a swim for half an hour or an hour daily is very beneficial to ducklings whilst in course of preparation for market. Only think how much more comfortable they must be when returned to a good bed of clean dry straw after a bath and a good feed. No one will ever convince me that ducks will do better, nor even as well, without water as with it. No, not even for killing.

Ducklings for the market can be and are bred in thousands by experienced breeders, who make it their sole business, and I know many of these breeders personally, who can and do make a very good living in England. There is no reason why our British farmers should not make considerable additions to their yearly income by breeding and rearing ducks for the market.

When we take into consideration the enormous quantity of ducks which are yearly imported into our markets by foreigners, I think I am justified in saying that there is not much fear of our markets becoming glutted with home productions, for as a rule English people are able to discern the difference between home and foreign productions, even in the shape of articles of food, and a good home-fed duckling is always in demand at a fair good price, at least that is my experience. I have never any difficulty in disposing of my wasters, when killed and dressed, at a very good price, and if I could not do better by keeping ducks for exhibition, I have sufficient confidence to believe that I could make a fair good living by breeding and rearing ducklings for the market.

METHOD OF KILLING.

As ducks are to be killed for the table or market, it may be as well to point out the most merciful way of taking their lives, a point on which few concern themselves.

The most merciful way, and also the quickest, if skilfully managed, is to take hold of the legs and flight feathers with the right hand, and the head with the left hand, place the neck over your left thigh, and with a sudden jerk the neck becomes dislocated, the spinal cord ruptured, and the bird is at once made void of the sense of feeling. The same effect may also be produced by a sudden twist.

Another plan is to take a stick in the right hand, hold up the bird by the legs with the left hand, and strike the bird a smart blow at the back of the neck, about the second or third joint from the head. Death follows instantly; but the breaking of the neck with a momentary jerk, as already described, is certainly the quickest method, consquently the most merciful.

We keep these creatures for our own pleasure and profit, and kill them for our use, but even in this our last and often painful duty we should treat them with mercy.

HOW TO TREAT DUCKS WHEN MOULTING.

As a rule, ducks begin moulting in July, and are usually through the moult about September. The earlier this change takes place the better. The weather being more favourable to the birds, gives them more time to recover their strength, and enables them to commence laying earlier in the season. Moulting is a very critical time even for ducks, and many valuable specimens are lost during this course of nature.

The general condition of the system being considerably reduced, it is therefore necessary to assist nature, if you wish your ducks to have an early and successful moult, by attending to their comforts internally and externally. They should be kept warm and dry. Their house should be cleaned out daily, and a good bed of clean straw put down for them to lie on. Warmth, generous diet, and cleanliness, together with an increase of animal food or other stimulants, will have the desired effect.

Warmth may be secured in many ways. See that there are no draughts in the house, that it is watertight and kept clean and dry inside. Increase the supply of meat, and use that kind of meal most likely to give heat to the body—viz., oatmeal, barley meal, or buckwheat meal, mixed with Spratts Patent Poultry Meal scalded with boiling water. Now is the time for the use of a little spice of some kind or other occasionally. I am only a very poor doctor, and am not an advocate either of spice or medicines. To prevent is better than to cure. Therefore see that your ducks are warmly and comfortably housed, supply them with good sound food well cooked. See that their food and water dishes are regularly cleaned and a good supply of gravel within their reach. If you do this you will not be troubled with many sick birds.

PREPARING AND KEEPING DUCKS IN CONDITION FOR EXHIBITION.

Although a recent writer tells us that ducks require less preparation for exhibition than any other class of poultry, my experience has been of a very different character to this. It may safely be said that a really first-class pen of young ducks, about six or seven months old, require very little or no preparation for the show pen, for they will just be in fine plumage. If Aylesburys the bills will be perfection at this age. The question now arises how are we to preserve these beautiful flesh-coloured bills, and the beautiful bloom and condition of the plumage?

Fine condition goes a very long way in a show pen, and hides many little imperfections. A really good young duck in perfect health and condition only requires placing in a training pen for an hour or two daily for a week previous to the show at which you intend exhibiting it. Some people think it is a waste of time to put ducks into a training pen previous to sending them to an exhibition, and perhaps it would be so in the case of Pekins, for they cannot stand too erect in their pens. With Aylesburys, Rouens, and Cayugas, it is very different, upright carriage of their bodies being very objectionable.

Birds should be accustomed to the show pen before being sent to an exhibition, otherwise they will not show themselves to advantage. They should be trained to eat grain of some kind when in their pens at home, so that they will feed without trouble at the show, and not require cramming. If you allow the old ducks perfect liberty, and they are continually exposed to all kinds of

80 PREPARING AND KEEPING DUCKS IN CONDITION FOR EXHIBITION.

weather, you cannot expect them to remain in good show form, consequently it is much better to allow stock birds to stay at home until they have moulted. To keep ducks in show form it is necessary to use some precaution to keep them very clean both in the house and run. It is impossible to do this during the breeding season. Therefore, if you wish to breed from your exhibition ducks you must be prepared to sacrifice them as exhibition specimens for the season, for if you allow them perfect freedom they cannot be expected to recover until they have moulted.

During the time this change is being effected is the time when you may assist them to that state of perfection in which they appeared at their first show. This can be done by good feeding and housing. The food should now be of a better quality, giving a little fat or linseed in their food two or three times weekly during the moult. Some useful hints will be found in the chapter on "The Treatment of Ducks when Moulting."

If your ducks have commenced moulting and it is about the middle of July, I would advise you to assist their progress during this change, in order to be able to compete successfully with others, and to do so, the first part of the business is, to examine the house and see that it is thoroughly clean and dry underfoot. A good coating of lime-wash, with a little carbolic acid or naphtha added, will clear the place of insect-vermin inside, and a coat of tar outside will make it watertight, and more suitable to get your birds in condition for exhibition, they will rest and do all the better afterwards Your house, having undergone a thorough cleaning inside and out, is now ready for the reception of your ducks which are to be taken up for repairs. They have had perfect liberty for say six months. Their feathers are hard and brittle, in fact they are worn out, and nature has ordained that ducks should have a change of feathers annually. The outer skin of the bill also changes with the plumage, and in order that these new feathers and skin may be grown to a high state of perfection, the birds on which they are

PREPARING AND KEEPING DUCKS IN CONDITION FOR EXHIBITION.

grown must be well nourished and comfortably housed, and when such feathers, etc., are grown to perfection, they must be preserved in that state by artificial means. If they are allowed perfect liberty during their moult, and more particularly, immediately after their moult, they will very soon damage their plumage, and their bills will very soon assume a gross appearance. It is therefore necessary to keep them in partial confinement during and after moulting. They should be let out for an hour or two daily to have a wash and a swim, and also to remain in the open air for a reasonable time, so that the sap may be dried out of the new feathers, which in an Aylesbury will very often be yellowish during moulting, in which case they must have access to the water and be exposed to the air.

During the time your ducks are having their bath and exercise, clean the house thoroughly and give a good bed of fresh, clean straw. Should their bills be too high in colour (yellowish), or if any horny substance has grown on the sides of the upper mandible, take a sharp penknife and carefully remove all objectionable superficial matter, which may have accumulated during the breeding season. Care must be taken *not* to touch the inner membrane, and on no account draw blood; for if you do it will cause the birds great pain, and it might do serious injury, or even cause death, therefore I would advise all beginners to be careful and not make too free with the knife in this matter, but rather use a bit of the finest sandpaper. If you just take the rough off with your knife, and then rub gently with the sandpaper, you will soon be able to discern when it is time to stop. This operation, judiciously performed, gives nature assistance in the performance of this change of skin on the bill, which is quite as important as that of the plumage. You will find that this change requires time. After taking off all surplus matter, you may put a quantity of clean gravel into a large bowl, five or six inches deep, fill up with water, and put it into your ducks' houses, and instead of feeding your ducks in the usual way in the afternoons, give them a few handfuls of wheat, and throw it into the bowl in which you have already put gravel and water. Your birds will, as a matter of course, root about in the bowl for the wheat, and whilst doing so they are scouring and bleaching their bills in a much better fashion than you could do so for them.

My remarks on the superficial growth of matter on ducks' bills, and the removal thereof, are equally applicable to Aylesburys, Rouens, Pekins, and Cayugas, and should have special attention when moulting.

The plumage of all varieties of domestic ducks is no doubt preserved and the colour brightened by partial confinement, especially during a hot summer-day, when the sun would tan the bills of Aylesburys, and burn the bloom off the plumage of any kind of ducks, no matter what colour. If you wish to keep your ducks in condition for exhibition, you must observe cleanliness in every particular, both in the house and outside. Feed with English corn, there is *no* best, although I prefer *wheat*; but avoid Indian corn, especially for Aylesburys. I have no doubt—in fact I'm sure—Indian corn would improve the colour of Pekins. Still I do not like it as a food for either English cattle or birds, as it contains too much yellow fat. The use of Indian corn has caused the premature death of many a valuable specimen. Whilst you are keeping your pets in prison for your own personal pleasure and profit, do not forget that you have so confined them and that they are depending on you for their subsistence. Think of the absolute necessity of the grit in the water, the clean straw, and the bit of green stuff. I grow a large quantity of Malta and Coss lettuce for my ducks, and they seem to enjoy it very much, in fact no kind of green food seems to come wrong to them. It is really marvellous what can be done in the improvement of condition by careful and judicious treatment, without overfeeding or impairing the system. Many would-be-fanciers have often told me that they could do with ducks, but they find a difficulty in keeping their bills the proper colour, especially the Aylesburys. I have known people assert that they really believed it was an impossibility to keep them right for a month after they left Aylesbury. This is simply nonsense. There is certainly a bit of art in showing an old duck with a very fine bill, but there is little or no art in exhibiting a young one in perfect colour and condition. As for there being anything in the soil in and around Aylesbury which has an influence on either the colour of the bill or plumage is altogether untrue and misleading. I have bred, reared, and exhibited Aylesbury ducks, and sometimes shown them in creditable condition, and the ground over which they have run is heavy clay-soil, there is not one particle of chalk in it that I am aware of. If you get your stock and your eggs from breeders who keep only one pure variety and attend to my instructions in this chapter, you will find that there is not much difficulty in keeping even the white Aylesbury duck in perfect health and condition.

I think I have said sufficient about getting birds up to show form, unless I introduce the scissors and tweezers for the benefit of young fanciers who may be tempted to indulge in a bit of

TRIMMING.

Perhaps they will be as well left in the workbox, for fear some new beginner should be too bold and try his hand before he has made himself sufficiently well acquainted with every shade and shape of all the required points necessary in a Rouen or any other variety subject to freaks which do not meet with the approval of their owners. I do not know exactly how they are done myself, but this I do know: I have seen a Rouen drake with a broad white ring round his neck at one show, and at another show very shortly afterwards I have seen the self-same drake on the scene with a beautiful narrow white ring, just the thing the doctor ordered. A few white or rusty feathers also have a most accommodating habit of moulting out before the rest of the plumage. I once knew a really first-class duck of this variety which up to two years of age was very bad to beat. Now when this duck moulted in her third year, she thought she would dress rather different from the rest of her sisters. So she came out with a charming white ring, which completely encircled her neck. The owner, an old hand, scarcely fell in with the old lady's taste. So he altered it—how I do not know, but the next time I saw her the white ring had disappeared and her neck one uniform colour.

Two flagrant cases of trimming were discovered at one of our leading shows. It frequently happens that a Rouen drake will grow more or less white feathers just at the extremity of his armour-chain, almost forming a white line between the chain or grey colour, which really ought to run close up to the velvet-black that covers the lower extremity of the body. It appears that some of these objectionable white feathers grew in this particular part of the plumage of two drakes, and the breeder seeing that they were very good in all other points, and finding that he could not win with them at a small agricultural show, decided to remove the white feathers, which he did very clumsily by breaking them off, leaving the small quills about a quarter of an inch long, also leaving a gap in the plumage, just as though it had been struck with a whip-lash. One of the best judges failed to discover the fraud, and awarded first and second prizes to the very drakes referred to. Now it happened that a new beginner fell in love with the first prize drake, and claimed him at something between £14 and £20, paid his money, got his receipt, and in less than ten minutes after doing this was advised not to enter him for a coming show, for if he did he would certainly be disqualified.

This is what I call hard lines for new beginners. And if judges and committees will not act consistently and disqualify all cases of trimming, they are neither working to the interests of the breeder, exhibitors, or the exhibition at which such trimmed specimens may be exhibited.

EXHIBITING.

Exhibition hampers.—I have frequently seen ducks sent to shows in hampers which were a disgrace to the owner.

A good hamper is best and cheapest. I like to travel my ducks in a large hamper. This hamper is 48 by 24 inches and 18 inches deep. In this I take four single ducks, and as it is divided with canvas into four compartments the birds are kept much cleaner than they would be if put altogether.

A closed wicker hamper about two feet in diameter and 18 inches deep is manufactured in Aylesbury; this is a very useful one. Many exhibitors often complain that their hampers have been cut open at shows. If exhibitors would only send their birds in proper hampers and have the lids to fasten with a strong little strap, they would have less to complain of

Instead of sending their birds in proper baskets which may be easily opened, many send them in old tumble-down things scarcely able to keep canvas and willows together. Then they take a packing needle and sew the lid or top canvas down, so that when the pen-men want to pen the birds they have no other alternative but to cut the strings.

Whatever kind of hampers you use, let them be light, consistent with strength. Let your fastener be a strap, and have them well lined and a pocket for prize cards if you wish to have them in good condition.

TRANSIT.— If your ducks are to be successful in the exhibition pen, they must be shown to the best advantage. Your birds may look exceedingly well before they are despatched, but by the time they arrive at the show they may, and very often do, look quite different. They have very likely had a long journey, and probably a rough one.

There are some parcel porters quite as careful as an exhibitor would be himself, and their are others who handle your birds in such a rough, heartless manner, that I am often surprised they get to the show alive.

No wonder we hear so many complaints about birds being damaged when at exhibitions. I have seen them piled up on a hand-cart three or four in height. The porter takes hold of the cart, runs along the platform, the baskets being piled up too high, come in contact with a board or something overhead which we see on platforms of most stations Down comes one or more of the hampers bang on the platform. No one knows or cares whether the birds are damaged or not. Away they go perhaps into the parcel office to wait for the next van, which may be one of the ordinary parcel vans.

These parcel vans are all right for a lot of small hampers, such as bantams, pigeons, and rabbits, but they are really too small for the conveyance of large poultry baskets. This being the only conveyance to hand, and as the exhibits must be in the show by a fixed time, there is no other alternative. The parcel van is loaded, a few large hampers soon fill the body, after which large baskets are lifted up on the top of the van, sometimes by two men, oftener by one, who will take hold of a hamper by one end and half throw it up to the top of the van. It is my opinion there is more damage done to show birds by the rough usage of porters and carters when loading and unloading at stations and shows, than all other accidents put together. *There* is *not* sufficient care taken by these men. Railway companies are beginning to see that it is to their advantage to make more provision for exhibitors and their stock, and those of us who are in the habit of visiting many of the largest shows have no cause to complain of the accomodation provided for us by most companies. A real fancier is always anxious to see his birds safely in the show, also to see how they compare with other exhibits, often a bit anxious to see the prize card up. and then *very anxious* to see his birds safe home again. Exhibitors can do all this by going to the shows themselves. Take your birds with you and bring them back if at all practicable. If your ducks are worth exhibiting at all they are worth showing well.

We will now suppose they have arrived at the exhibition either one way or the other, and my advice is, to the new beginner, "Go to those shows within easy reach." Do not rush your birds indiscriminately all over the country unaccompanied. If you intend showing at the Crystal Palace. Dairy, Birmingham, Liverpool, etc., keep a pen or two specially for these shows. Do not show them anywhere till then. By this time your ducks will be at their very best, and the greater will be your chance of winning.

PENNING.—Here we are again on the field at a summer show Birds at shows have travelled some distance either by road or rail. Perhaps they were put in their basket the night previous. The exhibition pen is one of the ordinary wire pens, 24 by 24ins., and 27ins. high placed on bare boards, not even a bit of canvas thrown over the pens to shield your birds from the burning sun or, on the other hand, a pouring rain. The pens are arranged in single tier in the open air. You have taken your birds to this show, and of course you took with you a bit of waterproof, a spare horse-rug, or anything, which you knew would protect your birds from the inclemency of the weather. Having made your pen comfortable, by putting a cover on the top and a nice bed of hay or straw in the bottom, open your basket and just see that your ducks are clean. If their bills or legs are dirty, take a small sponge and wash them. A bit of good wheat will be very acceptable If you put two or three handfuls in the water-tin your ducks will eat it and look all the better. Now shut the pen door, get out of the way of the judge, and wait patiently for the awards. If you are fairly beaten, or otherwise, do not get excited, but coolly reckon the matter up and find out, if defeated, the *reason why*. Then go in for the remedy, which in nine cases out of ten will be *better birds*.

CRAMMING.

I have said nothing about the objectionable practice of cramming ducks at exhibitions. I could relate several very amusing incidents which have occurred in this most important (to some people) finishing touch. I remember there was a resolution passed at a council meeting in connection with one of our largest winter shows, that cramming should, if possible, be put down. It was subsequently decided to have the ducks in the place of exhibition a whole day prior to judging. After all these precautions, what do you think happened on the day of judgment? Why one of the Rouen drakes fell sick.

This was reported in the office by the attendant, who by the way has been the principal attendant at this show for many years. A little later on, the same attendant reported, drake, pen No. so-and-so, "dead." Another report, duck, pen No. so-and-so, "ill," which also belonged to the same exhibitor, and yet another report, duck, pen No. so-and-so, "dead." Would you think it possible that this very drake and duck were the property of the person who was so anxious to put a stop to cramming? I cannot say for certain that these two birds had been crammed, but it was generally believed they had. There is one fact well known to a few old fanciers, that these ducks actually vomited nearly two pounds of raw beef. Some said it was horseflesh. This is scarcely the way to put cramming down. I think I could suggest a much better way, which would be fair to all. I am now speaking of duck-exhibiting at first-class shows, such as the Dairy, Palace, Birmingham, Liverpool, etc. Ducks, like Carriers, Barbs, and Trumpeters, should be provided with two separate cans, one for food and the other for water. Ducks ought not to be fed with soft food before judging. We all know what dirty creatures they are. If some wheat were put into one of these cups, and water in the other, birds will be able to feed well without making themselves so very dirty. If this plan were adopted, many fanciers would never try to cram.

The only way to prevent such abominable practices is for one exhibitor to have a good feeling towards another, and have a general understanding that this disgusting practice shall not be indulged in or tolerated. There is a difference between cramming a bird to suffocation and just giving him four or five pellets to keep him in condition.

Many young birds will not eat at exhibitions, then it is sometimes necessary to give them a little food in this way. Ducks which have been subjected to cramming are of little or no use for stock purposes. "Do not cram if your birds will eat."

THE QUESTION OF PROFIT,

FIRST YEAR.

This is an item about which there are a variety of opinions However, in this case I am not going to attempt to show something unreasonable. I have no desire to mislead or deceive anyone, having no personal benefit to gain thereby. I shall not under-rate the cost of stock, or be foolish enough to say that I can keep a duck for a penny a week; neither will I over-estimate the value of your produce for the sake of showing a large profit. I will simply show what has actually been done and what can be repeated without a shadow of a doubt. I do not say that it would be wise on the part of a beginner to plunge headlong into duck breeding to try to make £50 a year straight off the mark. *By no means.* He may already have a fair knowledge of ducks, and also have my figures to work by; still before he can make £50 a year, it is necessary he should serve a short apprenticeship

THE QUESTION OF PROFIT, FIRST YEAR. 91

In showing how to begin, you will observe I start with one pen of stock-birds. In my article on breeding, you would no doubt notice that I had only one drake and three ducks, in which I said we ought to get sixty eggs from each duck, in all 180. This is a very moderate calculation. Some ducks will lay almost twice that number; others will scarcely get up to sixty. However, mine have averaged about sixty each for the last four years, that is reckoning old and young. From 180 eggs we ought to get 150 ducklings from stock-birds which are not exhibited. I will endeavour to show the beginner what profit he ought to make by breeding from this one pen of ducks during the first year of his apprenticeship. First of all I will take the cost of his house and stock:

	£	s.	d.
Duck-house	1	0	0
One Drake and three Ducks 20. each	4	0	0

Then comes the question of keep. I allow 2½d. per week for each duck; this includes straw for bedding. This allowance is quite sufficient to keep large ducks in good health and condition. We have four stock-ducks which require feeding for fifty-two weeks, allowing that in summer or autumn we begin. Four ducks at 2½d. per week each for fifty-two weeks will cost £2 3s 4d. Then assuming we have 150 ducklings (certainly not all at once), some of them will be killed before the others make their appearence, and in order that there may be no mistake, I will calculate what will be the cost of feeding the whole 150 up to nine weeks old. These I allow the same amount per week as the stock-ducks, for although they eat very little the first month, they have a way of taking kindly to their food at about five weeks old, and don't they put it out of sight quickly for the next three or four weeks! and it is best and cheapest to let them have food to their hearts' content. Do not think for a moment that I expect you to rear the whole of the 150. There will, or I am much mistaken, be some casualties for which I will afterwards make an allowance. The cost of keeping 150 ducklings from birth up to nine weeks old at 2½d. per week each will be £14 1s. 3d. I will now divide 150 ducklings into packs of ten as suggested in a former article. Ten will go into 150 fifteen times. Now out of each lot of ten I will send eight to market, which will average 3s. 6d. each, and if you are only in the market early enough in the spring the average will be very much higher. There will be no necessity for selling any of your ducklings for less than 3s. each even at the end of the season, for, being bred from good stock and well fed, they are very different articles of food to the little, half-starved things we so often see in our markets and in our fish and poultry shops.

Superior ducklings will always command good prices. Here we have fifteen times eight—viz., 120 sold for market at 3s. 6d. each = £21. I will now make an allowance for casualties. Allowing one duckling out of every ten has either sickened or met with some accident or another which has put an end to its existence. This will be fifteen, for which I get no return whatever. Then I have fifteen selected ones, which are the best birds out of each pack Fifteen such birds as these when about six months old, if in practical hands, would realise a good round sum; but as they are in the hands of a beginner, I must not estimate them at too high a figure. I will therefore value them at the price as the same owner's stock-birds, which is £1 each, so that fifteen selected birds are worth £15. Of course these fifteen ducklings have been kept from nine weeks old up to twenty-six weeks I have therefore to add seventeen weeks' keep to the nine weeks already charged for at 2½d. per week, which will be £2 13s. 1½d.

Then there is the breeding stock which, if not sold off at the end of the first year, must be allowed for as wear and tear. If we allow twenty-five per cent. for depreciation of stock, and use the breeders for three years, and then sell them off for killing, we should not be very far on the wrong side. We have now to make some allowance for sitting hens or the use of an incubator. If we say £1 3s. 2½d. it will be fairly good pay for the use of twenty-two "cluckers" or an incubator.

I have said nothing about rent of houses and runs. The heap of rich manure you will have at the end of the season, together with the bagsful of fine feathers, will pay for the use of any outbuilding you may have occupied with your youngsters; or if you had to put up a rough shed, the feathers and manure will pay good interest on the outlay. The stock-birds will do more good than harm on your croft, garden, or pasture.

Now as to the question of labour. This has no place in the expenses, for most small farmers, gardeners, cottagers, or small fanciers, have either sufficient spare time themselves, or have children who can do this kind of work without interfering with anything else, and a pursuit of this kind often keeps a man with little to do out of worse mischief, and at the end of the year puts something in his pocket for doing so. We will now have a look at figures and see how the expenses and income tally for one year.

EXPENSES.

	£	s.	d.
Feeding 4 breeding ducks for 52 weeks at 2½d. per week each	2	3	4
Feeding 150 ducklings for 9 weeks at 2½d. per week each	14	1	3
Feeding 15 ducklings from 9 weeks to 26 weeks at 2½d.	2	13	1½
Depreciation of breeding stock 25 per cent. on £4	1	0	0
Use of sitting hens or incubator	1	2	3½
Total	£21	0	0

INCOME.

	£	s.	d.
120 ducklings for market sold at 3/6 each	21	0	0
15 ducklings for exhibition or breeding. at 20/- each	15	0	0
15 casualties	0	0	0
Total income	£36	0	0
Total expenses	21	0	0
Nett profit	£15	0	0

SECOND YEAR.

Having finished your first year's apprenticeship with a nett profit of £15, you must now decide on your course of action for the second year's work. It will not be necessary to keep three times the quantity of breeding stock in order to make three times the amount of money you made last year, but you must certainly increase your stock.

I should not advise you to keep more than two pens during your second year. The way to start your second pen is to select three of the very best ducks out of the fifteen you have bred, and which *to you* are really worth more money than you would perhaps be able to sell them for. Your only extra expense will be a new house and another drake. Do not buy a new stock-drake until you have visited one or two shows, at which you may see one in the selling class (a young one, mind). Satisfy yourself that he is not related to your original stock, and also that he is quite as good, or even better, than the best drake you have bred. It would be of great assistance to you if you made just one entry in each class, viz , one drake in his class and one duck in her class. You would then be able to compare your birds side by side with those of experienced breeders. Theory is all right and indispensable, and, when assisted by practical experience, often attains the desired goal. By exhibiting your birds you advertise them in such a manner that no other mode of advertisement is equal to. Old exhibitors see them. These gentlemen are always open to buy a really good one of any kind, and will not grudge to give a good price. In addition to the old exhibitor, we have the general public, some of whom have quite as keen an eye for a "topper" as the oldest and best breeder. If you do not meet with a drake at the show, you will be able to form some idea of other breeders' stock by the birds they are showing, and you will not have much difficulty in diciding where to go to look for one, or who to write to. If you write for a bird, have him sent on approval, and once more, "be sure he is a young one." Keep your eye on the *marking ring.*" The end of November is quite late enough to mate up your stock-birds. By this time you will have bought a drake, and, of course, put him to your three young ducks. You will now proceed with your two pens of breeders the same way as last year, setting all the hens as directed. If broody hens are bad to get, which is often the case in winter and early spring, the next best thing to do is to invest in an incubator. As I said before, the "Westmeria" is my favourite; it is the safest and best machine I have ever come in contact with

Suppose your success in breeding is equal to last year, your eye is better trained to enable you to be more accurate in your selection of the best birds, and if your selection of stock-birds has been a judicious one, it will not be unreasonable to expect two good birds out of every ten from your second pen of ducks. I will now see what profit is derived from your No. 2 pen :—

EXPENSES.

	£	s.	d.
Feeding 4 stock ducks for 52 weeks at 2½d. per week each.	2	3	4
Feeding 150 ducklings for 9 weeks at 2½d. per week each.	14	1	3
Feeding 30 ducklings from 9 weeks to 26 weeks at 2½d...	5	6	3
Depreciation of breeding stock, 25 per cent. on £5 (this is allowing you gave £2 for your new stock drake)...	1	5	0
Use of sitting hens, or wear and tear of an incubator......	1	2	0
Allow for entry fees and expenses at two shows	4	9	8
Expenses............£28	7	6	

INCOME.

105 ducklings sent to market, at 3s. 6d. each...............	18	7	6
30 ducklings for exhibition or stock purposes, at 20s each.	30	0	0
15 casualties...	0	0	0
Income £48	7	6	
Expenses 28	7	6	
Profit......£20	0	0	
The result from No. 1 pen, same as last year 15	0	0	
Profit for the year...£35	0	0	

THIRD YEAR.

Having served two years as an apprentice, I will now ask you to commence work as a fully-fledged journeyman. First of all clear out your first pen of breeding ducks, which at two years old should be good enough to sell at 10s. each, and with the money you get for these purchase another drake, and put this to three of the very best ducks you bred last year. Having done so, you have now two really first-class breeding pens, and as your six ducks are the cream of three hundred, and mated with two drakes purchased from two of the best breeders, proceed as usual throughout the next season, and if you have succeeded in accomplishing the not-too-difficult task I have set you during the last two years, you will make £50 profit this year, and make this sum easily. If you are successful as an exhibitor, I shall not be going too far if I say £20 over this sum.

Personal attention, judicious selection, and the early despatch of wasters will carry you to the top of the tree this season if you only follow my instructions. The cause of so many failures is the lack of *personal attention*, and bad judgment. Bring to bear a little common-sense, and do not be disheartened by a few disappointments either in breeding or exhibiting. Give a "long pull and a strong pull." Be thoroughly determined to accomplish your object, and then you *will* succeed.

I will not trouble you further with instructions, but proceed to show how *I* can make £50 a year by keeping and breeding ducks. I rear annually as near 300 ducklings as possible, I kill all wasters as soon as ever they are fit, and I keep the selected ones either for exhibition, stock purposes, or for sale. If I had 300 ducklings this season, bred as I have advised others to breed, I should kill for market about two-thirds of the 300, which would leave me 100, barring accidents, which would realise a fairly good sum, perhaps more than the new beginner would be able to obtain. Of course I should not sell the whole lot, but reserve about six for my own use, viz., stock and exhibition birds. These would *not* be six of the *worst*.

This is the way I make £50 a year by breeding ducks, saying nothing at all about keeping a larger breeding stock from which eggs are sold, which, by the way, sometimes add considerably to the income of a duck-breeder. Neither have I said anything about the profit to be derived from exhibiting, a subject on which I could give a fairly good account.

Now after all this exhibition of my own skill as a duck-breeder, it will still be necessary for me to show by figures how it is possible for a new beginner to make £50 a year. Here we have six ducks and two drakes. The ducks are calculated to lay sixty eggs each in the season, which equals 360, from which we should have 300 ducklings. Two-thirds, which is 200, are sold for market at 3s. 6d. each when they are nine weeks old. Thirty die from some cause or other, and the remaining seventy selected ones are kept until they are six months old, when the value will be on an average 20s. each. A few will be worth considerably more than this price, others perhaps a trifle less. All this being as written, the expenses, income, and profit will be as follows:—

THE QUESTION OF PROFIT, THIRD YEAR

EXPENSES.

	£	s.	d.
Feeding 8 stock ducks for 52 weeks at 2½d. each per week.	4	6	8
Feeding 230 ducklings for 9 weeks at 2½d. each per week.	21	11	6
Feeding 70 ducklings for 26 weeks at 2½d. each per week.	18	19	2
Use of hens or incubator	2	0	0
Allowance for entrance fees and expenses at shows	5	12	11
Depreciation of stock, 25 per cent. on £10	2	10	0
Total expenses	£55	0	0

INCOME.

	£	s.	d.
200 ducklings sold for market at 3s. 6d. each	35	0	0
30 casualties	0	0	0
70 selected ones sold or in stock at 6 months, at 20s. each.	70	0	0
Total Income	£105	0	0
Total Expenses	55	0	0
Nett Profit	£50	0	0

GEESE.

Like the duck, the common goose ceases to be strictly monogamus when kept in a state of domestication, yet we often find that a gander will pay more attention to some particular goose than to the others which may be mated with him. The instinct is not altogether obliterated, for there is generally a reigning sultana. Therefore, it is not a good plan to mate a large number of geese with one gander, three being quite as many as is safe for breeding purposes.

The oldest and best authorities all agree that the "Wild Gray Lag Goose" is the original parent of all our domesticated varieties. At what period and by whom the goose was reclaimed it is difficult to say, but that it has descended from the "Wild Gray Lag Goose" cannot be doubted. Neither can there be any doubt that the common goose was found in a state of domestication in England by Cæsar, and long before his time the goose was widely spread in other countries. The "Gray Lag" is an irregular migrator, sometimes visiting the central and eastern parts of Europe, Asia, and Africa. Formerly it was a prominent resident in England, and took up its quarters in Lincolnshire and Cambridgeshire, and other fenny districts, where it bred freely.

In consequence of the vast improvements which have been made in agriculture during the last century, the advancement in commercial enterprise, and the great increase of human population, a great change has taken place. The fens and marshes have been drained; extensive marshes frequently under water have also been converted into rich farms, and are now devoted to the plough. Consequently, the "Gray Lag" is no longer a permanent resident in our island. Still a few stragglers sometimes pay us a visit during the winter, and remain with us for a time; but the gun of the sportsman very soon brings them down, or disturbs them in such a manner that they do not remain with us very long. They are naturally very shy, and are continually on the alert, even whilst feeding. Sentinels are on guard, and on receipt of the signal of alarm, the whole flock instantly rise on the wing. This precaution is observed night and day, and this instinct appears to be retained in all our domesticated varieties. A set of geese are quite as good guards as a watch dog, for they will raise an alarm at the sound of footsteps, or at the least suspicion of the presence of a "stranger within the gates."

From the "Gray Lag Goose" I will now turn to its domesticated descendants, which vary to a considerable extent in colour and symmetry. We have the common goose, well known throughout the British Islands.

I remember when, not very long ago, I traversed extensive commons in different parts of England and Scotland, seeing thousands of these common geese, the flocks of the respective owners generally keeping well together, and should they by chance mingle with those of other owners', they usually separated themselves towards evening. Each flock was driven home by a boy called a "gozzard" (goose-herd). The flocks of goslings were brought up together and formed a united band. Each flock was marked sometimes by dabs of paint of various colours, and sometimes by marks punched in the webs of the feet. It was only on very rare occasions that these marks were of service, for being brought up together, they appear bound by the ties of habit to remain so.

During the last fifty years many of these large commons have either been enclosed or very greatly circumscribed, and the number of geese which were kept by the people around, who had the right of grazing, has been reduced accordingly. Still there are thousands of geese reared annually in England, and there is room for thousands more.

Lincolnshire is still famous for rearing common geese. Pennant tells us "The geese in his time were kept in multitudes in the fens of Lincolnshire, a single person frequently having a thousand old ones, each of which reared on an average seven young ones, so that towards the end of the season the owner became possessed of about 8,000 geese." "The stock geese, says Pennant, "were plucked five times a year for the sake of the feathers." Let us hope it is not so now, and in fact I do not think it is, especially when we take into consideration the quantity of feathers we import from other countries, no less than £105,526 being paid to foreigners in the year 1891 for feathers for beds. Independent of this fact, we must not tolerate the practice of plucking live geese, for it is a barbarous custom, and should be thoroughly abolished. I know it has been the death of many farmers' geese in my time, and mortality must have been considerably greater in Pennant's time

Lincolnshire, Suffolk, Norfolk, and Berks, send the best geese to the London markets. Geese should be housed apart from any other kind of poultry, and have large houses or sheds to sleep in, and always be provided with a clean dry bed of straw to lie upon. A grass run is absolutely necessary for rearing goslings successfully, and a convenient pond or stream of water is conducive to their health and well-being, although not so necessary for the successful rearing of goslings as it is for stock birds. Owners of geese should not suppose that grass is sufficient to keep them in perfect health and condition, for they require in addition a daily supply of corn, such as oats, wheat, barley, or maize. Without this they cannot do credit to their owners and grow to any considerable size.

Hundreds of young geese which might have been reared on the commons, and on poor land, have pined away and died for want of sufficient nourishment, diseases of various kinds have been manifest, and then aggravated by cold and wet, their impoverished systems being destitute of stamina, they droop and die one after another, until that which was a fine flock in the spring is almost annihilated by the autumn. Thousands of goslings are annually brought in from the country by experienced feeders in the vicinity of London and Belfast. The management of goslings by these wholesale feeders is simply what it *must be* if profit is a consideration. The whole business is conducted with regularity, cleanliness, and punctuality.

The great object in preparing geese for market is to do it in as short a time as possible. Unremitting attention must be paid to their comforts and requirements, they must be supplied with proper food and at proper intervals. Water and exercise must also have due attention when preparing goslings for the market, but I shall have more to say on this subject in another chapter.

The common goose varies in colour from the grey to the grey-and-white saddle-back, and then to the perfect white. There is no doubt in my mind that our present domesticated sub-varieties are the outcome of man's skill in breeding. Our magnificent exhibition specimens have, by the art of man, been brought to a high state of perfection, and that has been done by careful and judicious selection of stock for a large number of years. All our domesticated geese, including the common farm-yard goose, the loose-skinned Toulouse, and the tight-feathered Embden are of one common origin, the "Wild Gray Lag Goose."

Toulouse Gander, bred by Henry Digby, winner of first at the Crystal Palace; first and cup, Birmingham; first and cup, Liverpool; and many other prizes.

THE TOULOUSE GOOSE.

"Toulouse Geese" are my favourites. I have bred and exhibited them more successfully than any other variety, and my idea of the standard of perfection in this variety is identical with that of D. Bragg, Esq., Southwaite Hall, near Carlisle, who wrote an excellent article on this subject, and which appeared in the Christmas number of the *Fanciers' Gazette*, 1890. From this article I have extracted several remarks, by the kind permission of the writer, whom I know to be one of the best living authorities on geese, also one of our best and most popular judges of waterfowl. These remarks are chiefly confined to the standard of excellence for exhibition and will be found useful to breeders of geese. I do not believe in feeding my geese up to extraordinary weights for exhibition. It is one of the most fatal mistakes that can possibly be made. It is the absolute ruin of a good bird, reducing such to barrenness, and often resulting in premature death.

It is the proper course to get the frame as large as possible, and with it a sound constitution in order to carry a large weight of flesh consistent with the size of the frame. Too much fat, especially internally, is most detrimental and dangerous to the lives of your geese. I prefer my geese being judged by appearance rather than by weight. My pair of Toulouse geese which won at all our leading shows during the year 1888-9 weighed forty-eight pounds, and my greatest difficulty with them was to keep the weight down. So great was their disposition to fatten that it was very rarely they got anything but grass in summer and a little bran and thirds in the winter. The accompanying sketch is a true representation of the gander in question.

At the Dairy Show, I claimed Miss L. Picken's gander, which won the first prize and the Association medal, 1890. This gander is, in my idea, perfection in style, colour, and build. A capital sketch of this bird appeared in the Christmas number of the *Fanciers' Gazette*, 1890. I am informed on reliable authority that he is one of Mr. Bragg's strain. But be that as it may, he is just the type of a Toulouse gander from which I shall draw up my standard of perfection, assisted by Mr. Bragg. No doubt this is the modern and improved type, and has met with general approval, and deservedly so. The chief improvement in the form and symmetry of the Toulouse is a prominent and deep breast, resembling as much as possible that of a first-class Rouen or Aylesbury duck.

The keel should be perfectly straight from stem to paunch, where it divides evenly and increases in width to stern, yet forms a straight underline. All indentations in front of thighs to be well filled out, otherwise the full, deep breast will resemble a loose crop. Stern to be heavy and wide. In contour, a full rising sweep from paunch to a high-carried-spreading tail; shoulders, broad; narrow shoulders are a general failing; neck, medium length and thick; the head should be strong and massive, in profile, an uniform sweep, or nearly so, from point of bill to back of skull, resembling in formation the head of an Antwerp pigeon. Throat, moderately gulleted, which is uniform with the loose-skinned body of the Toulouse. A side view of the bird should appear almost square, and viewed from behind, an oblong square. The same description answers for both goose and gander of this variety, for it is sometimes difficult at exhibitions to say which is the goose and which the gander, but the sex should always be *bonâ fide*.

Colour of bill and legs, orange. Any white round the base of the bill is very objectionable, and should be bred out if possible; it comes with age in some strains, and I do not remember seeing many really good old Toulouse geese without this fault. I have known them to be pulled out, and in more than one case dyed; therefore I would not pass or even score many points against a Toulouse with this defect apparent.

Neck feathers, dark grey. Back, wings, and thighs, dark steel-grey, each feather laced with an almost white edging. The breast should be a sound grey colour, without lacing, and to descend as uniform as possible through the keel. Toulouse are very liable to run too light in colour below the breast. The colour does naturally run lighter from the breast to the legs, but when viewed in front, little or no white should be seen. The stern, from the legs to the tail, should be white. Tail, white with a broad band of grey across the centre. The wing-flights, a self-coloured dark grey. Eyes, large, bright, and dark. Twenty-eight pounds is a good weight for a matured gander, and twenty pounds for a goose in fair store condition. I do not wish to have them more than these weights.

The Author's Ideal Toulouse Gander. The Dairy winner of 1890 improved.

SCHEDULE FOR JUDGING TOULOUSE GEESE.

VALUE OF POINTS IN EITHER SEX.

		Points
Head,	Strong and massive.	
Eye,	Dark and full.	
Bill, and Throat.	Orange, strong and well set, in a uniform sweep, or nearly so, from point of bill to back of skull. Well gulleted.	15
Neck.	Good length and thickness, dark-grey colour	5
Breast and Keel	Prominent, deep and full, a sound grey colour, shading a trifle lighter to thighs. Perfectly straight from stem to paunch, increasing in width to stern, forming a straight underline. Colour same as breast.	10
Colour and Marking.	Back, wings, and thighs: dark-steel grey, each feather laced with an almost white edging. Flights: a sound grey without white.	10
Tail, Stern and Paunch.	Carried high and well spread, white, with a broad band of grey across the centre. Heavy and wide. Colour: white	10
Size	As large as possible. A good well-matured gander should weigh 28 to 30 lbs., and a goose 20 to 22 lbs.	20
Symmetry	Massive head, carried majestically on a strong, well-formed neck. Body: long, broad, and deep. Shoulders: broad. Breast and keel: deep, forming a straight underline. Paunch and stern: a full, rising sweep to tail.	15
Legs and Feet.	Orange, and very strong in bone.	5
Health and Condition	Plumage: full, bright, and glossy, and in perfect order throughout. Eye: bold and clear. Body: heavy, but not broken down.	10
	Total number of Points	100

DISQUALIFICATIONS.—Crooked back, wry tail, slipped or cut wings, or any other bodily deformity, and patches of black or white amongst the grey plumage. The sex should also be bonâ-fide.

Embden Gander, winner at the Dairy and Birmingham, now the property of Henry Digby, Author of this work.

THE EMBDEN GOOSE.

THIS handsome and useful variety of geese have undoubtedly been neglected during the last few years, although they muster fairly well when separate classes are provided for them. But where are the Embden fanciers? And echo answers *where*?

Some time since one fancier wrote to the *Stock-keeper* complaining that all varieties of geese were jumbled together at the Crystal Palace Show. When I appealed to Embden fanciers to give their names and support to provide separate classes for Embdens at the next Palace Show, I failed to find that writer's name amongst my list of supporters. I may say that the Palace Committee are quite willing to give classes for Embden geese if fanciers will support them; but of course, they must have some guarantee that the classes will fill or pay in some other way.

It is quite true what "Poor Embden" says in his letter. "The superiority of the Embden as a table bird is unquestionable, and the encouragement it receives at the hands of 99 judges out of 100 is most disappointing." Some judges do not understand the merits of the birds sufficiently well to place them over their more bulky-looking relatives, the Toulouse. The Embdens are undoubtedly excellent table birds, prolific layers, and good breeders, and when we are able to record the sale of a pair of this variety for the handsome sum of £50, it is enough, or ought to be, to induce farmers and others who have convenience to keep a few geese to go in for breeding Embdens, especially when I tell you on reliable authority that the pair above referred to weighed 53lbs. 6ozs., so that if they had not been up to the standard of perfection for exhibition, they would have realised a fair price in the market for table purposes.

I quite agree with Mr. Bragg, who is one of the most extensive breeders, also one of the best judges of this variety, as to the required distinction between the Embden and the Toulouse. If other judges would only act consistently they would eventually put a stop to the cross between the two, at all events for exhibition purposes.

Embdens should have long broad breasts, with as little indication of keel as possible; shoulders, broad; long straight back; body, broad and thick, and as near round as possible; a long swan-like neck is a characteristic and an important set off to an Embden.

The head should be long and straight, avoiding as much as possible a "dished" upper mandible. The throat should be clean without the slightest indication of gullet. I would not disqualify an otherwise good bird if even it did show a bit of gullet. The tail should be carried close and straight out; bill and legs should be a rich orange colour; eyes, light blue; colour of body throughout a spotless white; carriage more sprightly and upright than the Toulouse, thus giving the bird a very smart and defiant appearance.

Embdens frequently weigh heavier than the Toulouse, although they really do not appear to be so large. They are, or should be, very hard and tight in feather, something like the Game cock, whereas the Toulouse is more like the Cochin, and geese should be judged quite as distinctly as the two varieties of fowls just named. A well-matured Embden gander will turn the scale at 30lbs., and a goose at 22lbs.

SCHEDULE FOR JUDGING EMBDEN GEESE.

VALUE OF POINTS IN EITHER SEX.

		Points
Head, Bill, Eyes	General appearance: long and straight. Rich orange. Full and light blue	12
Neck	Long and swan-like.	10
Breast	Broad and solid, with as little indication of keel as possible. (One judge says, without any indication of keel).	20
Plumage	A spotless white throughout.	10
Size	As large as possible. A well-matured gander should weigh from 30 to 34 lbs., and a goose from 20 to 22 lbs.	20
Symmetry	A long, straight head, well carried on a long, swan-like neck. Breast: broad and solid. Shoulders: broad. Back, straight. Frame: long. Tail carried well out and close. Paunch: deep. Stern: broad.	12
Legs and Feet	Bright orange, and very strong in bone.	6
Health and Condition	Plumage: compact, hard, bright, and glossy. Eye: full and bright. Body: heavy; and general appearance lively and defiant.	10
	Total number of points	100

DISQUALIFICATIONS.—Crooked back, wry tail, or any other deformity. Plumage other than white. Misrepresentation of sex.

BREEDING AND REARING GEESE FOR EXHIBITION.

In breeding goslings for exhibition it is very important to make judicious selection of breeding stock, which should consist of a gander and two or three geese. They should be of mature age (not less than two years old), and the gander should not be too closely related to the geese. In fact it is much better to use stock birds which have been bred from two distinct families. Therefore, if you are beginning a strain, I would advise you to purchase your geese from one breeder and your gander from another, although a few breeders, I have no doubt, could supply you with both unrelated. Still it is the safest and best plan to go to separate yards for your stock birds. If you find your geese prolific layers and good breeders, throwing a fair percentage of good birds, I would advise you to keep that pen of geese until you find that they do not lay more than twelve or fifteen eggs each in a season.

Geese vary considerably in their laying and reproductive powers—some living and doing well to a great age, ten years being no uncommon time for them to live and do well—that is, if they have been used solely for breeding purposes and have not been overfed or knocked about the country from show to show.

If you feed them on stimulating food in order to get extraordinary size and condition, you will at the same time undermine their constitutions and shorten the period of their natural lives. In situations adapted for breeding geese there is no doubt they are profitable; but to do well they require access to water and a grass run, which is absolutely necessary to successful breeding. When kept under such circumstances, they require very little hand feeding, a good handful of wheat, barley, or oats to each bird morning and evening being quite sufficient during the breeding season. They should have an outhouse to themselves, plentifully supplied with straw with which to make their nests on the floor. There may be two or three large American bacon boxes placed in the corners on their side, or a few partitions fixed so that the geese may (if they choose) make their nests and lay in them without being disturbed by each other. Geese commence laying early in February if the season is a mild one and the breeding stock have been fairly well nourished during the winter.

The Toulouse are the most prolific layers, but they are only indifferent sitters and mothers. I have owned individual birds of this variety which have laid no less than forty-two eggs in one season, and several others which have laid between thirty and forty. The Embdens do not lay quite so many eggs as do the Toulouse, but they are considerably better sitters and mothers.

The wild gray lag goose lays from five to eight eggs; the tame common goose from thirteen to eighteen. High feeding, careful selection, and other comforts bestowed on our high-class geese have increased their prolificacy to a very great extent, and it is not at all an unusual occurrence at the present time for a single goose to lay thirty eggs in a season.

A common goose will frequently lay a second time if even she is allowed to sit, and bring up a brood of goslings. I prefer sitting my goose eggs under large hens, giving each four or five, according to the size of the hen. I find hens much safer than geese for incubation. I have occasionally hatched goose eggs in the "Westmeria Incubator," but I would advise beginners to use hens in preference to either geese or incubators.

The period of incubation is thirty days, if the goose or hen sits well, and the eggs are all fresh when put down; but it may, and frequently does, extend a day or two over that period. If the eggs are entrusted to a goose, the house in which she is sat should be so arranged that the bird can have entrance at all times of the day, for the sitting goose generally leaves her nest once a day, when she should be fed with hard corn. It is a good plan to leave some corn in a bowl of water close by her, so that she may be able to feed at any time she may leave the nest, and it is desirable that she should have access to water at this time. Goose eggs should always be set on the ground or in large boxes half filled with moist soil, so that when the heat of the bird's body which is sitting on them has warmed the nest through (which should be made of hay) it will draw sufficient moisture from the soil, to soften the hard shell and the inside membrane, and so enable the gosling to extricate itself from its prison without the risk of being shell-bound. Should the season be a dry one, the earth or soil under and around the nest should be freely watered with warm water two or three times during the last fortnight of incubation. I never sprinkle water on to the eggs; there is often great danger in such a practice. It is much safer to water the ground all around the nest, and even the nest itself may be damped with advantage but the eggs should *not* have water sprinkled directly upon them. The process of watering the nest is best done at night, when all is quiet, as the sitting hen is then less likely to be disturbed, and the work can be done efficiently. A quart of warm water is not too much for each nest, especially if the nest is properly made. It is not desirable to interfere with the goslings at the period of hatching, further than to remove the shells and any other refuse which may be in the nest. They should be allowed to remain in the nest for at least twenty-four hours after hatching before being disturbed or fed. Some people take them away from the mother as soon as they are hatched until the whole lot are hatched out. I never indulge in any such unnatural practices, but I allow them to remain in the nest under the hen or goose who has hatched them for a whole day and a night after the appear-

BREEDING AND REARING GEESE FOR EXHIBITION. 117

ance of the first gosling. By this time all that are of any good will be hatched and are being nursed in such a manner that cannot possibly be equalled by any amount of artificial heat. The next day they may be removed to the shed or coop provided for them and may be supplied with their first meal and a bit of cut grass. Many people seem to think that goslings are very little trouble and expense as compared with other domestic poultry, but my experience is somewhat different. I find it necessary to use the best and most nutritious foods for very young goslings, that is, if I wish them to attain to a very large size, or arrive at a high state of perfection; and as I am trying to show beginners how to rear goslings for exhibition, I will endeavour to explain the ways and means by which I have succeeded during a long and successful experience. First, do not be misled by writers who tell you that goslings are the easiest creatures in the world to rear. Certainly they can and do exist under circumstances which would prove fatal to turkeys or many other varieties of poultry, but it should be remembered that to exist is one thing, but to *live well* is another. Prizes are not won now-a-days by animals and birds of only ordinary merit. There are now so many skilled breeders and exhibitors of most varieties that we must have something extraordinary to be successful, and in order to rear our birds to perfection we must use all fair and legitimate means to get this extraordinary size, and that can only be attained by a system of judicious feeding and assiduous care and attention in every particular. It is also essential they should have an abundance of fresh air, exercise, and pure water to drink and swim in. A good house well bedded with straw to lie upon at nights is another important necessity. If goslings get a good start they seldom look behind them, and they grow amazingly after the first week or two. I have had goslings during the last few years which, when only three weeks old, I found it impossible to get the marking rings over their feet. I have been surprised at the amount of bone they have made in so short a time.

My first feed consists of hard boiled egg, a bit of suet, and a few bread crumbs chopped up together. Of course there is, or ought to be moderation in the supply of such highly nutritious food. Eggs and bread are only necessary for the first three or four days. A liberal supply of fine sharp grit and green food of some kind is absolutely necessary from the first. This and sufficient clean water to drink during the first week or ten days will be all they require, and after that a swim will do them good; but do not drive them into the water, let them take to it on their own accord. When the goslings are three or four days old diminish the quantity of eggs or discontinue their use altogether. I prefer giving for six or eight days a bit of good, sound wheat, a few groats, or a bit of tip-top barley once or twice daily affords a healthy change. Now, as size is a great consideration, it will be beneficial to give goslings that kind of food containing a large percentage of bone-making material. Fine sharps or middlings, as it is often called, should be used along with the biscuit meal. The advantages of this method of feeding are many, especially if such a method is supplemented by good pasture. It is a mistake to suppose that goslings will thrive as well on poor, coarse land as they will do on rich, cultivated soil.

BREEDING GEESE FOR THE TABLE OR MARKET PURPOSES ONLY.

In breeding goslings simply for the table or market there will be no necessity for spending much money at the commencement. Geese which have been forced to an extraordinary size for the sake of winning prizes are neither necessary nor desirable for the production of goslings for the market. Now, although it is exceedingly undesirable to have recourse to the over large exhibition specimens for this purpose, it is most important that a proper start should be made, and that suitable birds should be selected in order to produce the best results

BREEDING GEESE FOR THE TABLE OR MARKET PURPOSES ONLY.

These can only be obtained by the use of strong, healthy, well-matured parents, which should be of medium size.

If you prefer breeding from a cross, let that cross be between two pure varieties. I would strongly advise you to use an Embden gander and Toulouse goose, or *vice versa*, but the Toulouse being very much more prolific layers than the Embdens, frequently lay double the number of eggs in a season; consequently they are the best and most profitable to keep for this purpose; and as they do not evince a desire to sit nearly so soon as the Embdens, their eggs may be set under hens, and double the quantity of goslings may be hatched and reared. Let me advise those who have already a good class of common saddle-backs (grey-and-white) geese to purchase a Toulouse or Embden gander to run with them, for I am sure the result would be very satisfactory. Either would increase the size and prolificacy of the offspring. Spanish or Canadian ganders may be used for crossing with other varieties, and such crosses are generally very successful.

It should be remembered that the heavier your goslings weigh, the greater will be the price in proportion. For instance, a well-fed gosling, weighing 20 lbs., will realise threepence per pound more than one weighing only 10 lbs.

Extraordinary good home-fed goslings always realise a remunerative price. Where there is good convenience for keeping geese, and they can obtain a large proportion of their food on the pastures and on the stubble, there is no doubt that they can be made to pay a good return for the money spent and the trouble bestowed upon them. The stock birds can be kept in a productive condition much longer than any other variety of poultry; in fact they have been known to breed well at fifteen and up to twenty years of age.

Old ganders are sometimes very pugnacious and ill-tempered, and they are often very dangerous where there are children, especially during the breeding season, therefore it is well to keep a sharp eye on them during this period, or they may do serious damage to unsuspecting children. It is best to give these old ill-tempered birds a wide berth, or to keep them in an enclosure where the fence is sufficiently good to keep them within their bounds.

An idea prevails with many farmers that any kind of light grain or the poorest of pasturage will do for geese and goslings. A greater mistake cannot be made; and those who breed and feed on a large scale know better, and invariably make it a rule to use the best They "Let the flock's good feed be the master's heed. What at first he may cast will be doubled at last." The *Lincolnshire Chronicle* for December, 1845, states that Mr. R. Fuller, a poulterer of Boston, killed in one week for the London Christmas market 2,400 geese, 1,000 ducks, and 500 turkeys, which weighed altogether upwards of twenty tons. Thousands of goslings are still sent from various parts of the country to persons engaged in the business of preparing them for the market, and this business is carried on profitably, and to a very great extent in the vicinity of London and Belfast.

The management of the birds by the great feeders is carried on as it were by machinery; with such regularity and system, cleanliness and punctuality prevails, and as soon as ever the birds are fit they are killed and marketed. Well, now, if it is profitable to breed goslings and sell them to these feeders, who also have to make a profit out of them by the time they are ready for consumption, I ask whether it would not pay the breeder considerably better to keep his goslings until they were ready for the market, and then send them direct to the nearest or best market? Or, better still, where practicable, to sell them direct to the consumer, and by doing so reap the whole benefit of your own industry, and save the profits which the middlemen under the present circumstances gain?

I am quite aware that it is not always possible to avoid dealing through middlemen, but such dealings should be avoided as far as possible. Green geese are sent from the

BREEDING GEESE FOR THE TABLE OR MARKET PURPOSES ONLY.

southern counties to the London market very early in the season, when they realise good prices. They are well fed from the shell until they are about four months old, at which time they pay the best return. The great secret of rearing goslings profitably for the market is to keep them going by feeding them well all the way from the shell to the market, and as soon as ever they are ready let them go, and do not keep them to eat their heads off. If proper attention and good food is bestowed upon them from the first there will be no necessity for any particular course of fattening, for when they are big enough they will be fat enough. Good fat goslings are always good to sell, so that there need be no excuse for keeping them till Christmas. Farmers having stubble and a quantity of light corn can profitably afford to run goslings on to the festive season. Small keepers, who, I will suppose, have no stubble, and have nearly all to buy for their birds, must act very differently. They must make the most of their goslings in the shortest time possible, or the balance will most assuredly be on the wrong side. Goslings, as a rule, have good appetites, and are capable of consuming a large quantity of food, hence my reason for pushing them on to the very end without a break.

Goose breeding and rearing may be carried on successfully and profitably by many farmers without interfering with other branches of farming industry. This can only be effected by the observance of a few simple rules. In the first place, judicious selection of breeding stock is very important, for it is only large birds that will pay for the trouble and expense of rearing, and as they eat very little more than small ones the profit is considerably greater. Secondly, goslings intended for an early market must be as well fed and as comfortably housed as any other class of live stock. Thirdly, they should be killed and sold as soon as they are in the pink of condition for the first time. If they are allowed to pass this stage, I question very much whether they will ever pay as well, if even they weigh a few pounds heavier. Last, but not the least rule to consider, is the best way to kill and market your goslings.

There is a great diversity of opinion as to the best way of killing geese. Some people are neither quick nor merciful when performing this operation, and I think it needless for me to describe many of the ways and means adopted of taking the lives of geese, for some of them are exceedingly cruel. The best way of killing a goose is "first catch it," afterwards tie the legs together with a piece of strong twine, then cross the wings to prevent it from knocking about. Drive a strong nail or have a hook fixed in some convenient situation in the backyard or in an outhouse. Hang the goose up by the legs, the head will then, as a matter of course, be downwards; take a short thick stick in your right hand, and strike a smart but not heavy blow on the back of the head, which will stun the bird for the time. Have a sharp penknife in readiness, and the moment you have struck the blow with the stick take up your penknife in your right hand, hold the head of the goose with the left, then at once proceed to stick the bird through the neck immediately behind the extremity of the lower jaw. This should be done with one thrust of the knife, and should cut the main artery, or jugular vein, just in the same way as a butcher would stick a sheep.

Geese killed in this manner frequently die without a struggle, and so far as my experience goes, I have found it the most expeditious and consequently the most merciful method of taking their lives. Now that a goose has been slain, the next thing to be done is to pluck it. This is best done whilst the body is still warm, therefore no time should be lost, but as soon as ever life is extinct plucking should be commenced and finished as soon as possible The feathers should be selected as they are plucked, *i.e.*, all strong quills and secondaries should be put in a receptacle by themselves, and all the fine feathers and down in another, for good goose feathers have still a certain market value. Therefore they should be carefully selected and preserved whilst plucking is proceeding. Should the goose in question be destined for market, it should be put into shape as soon as it is plucked.

The legs and feet should be twisted on to the back and the wings on to the shoulders. The bird should then be laid out in a cold place, with its breast upon a cold stone, marble, or granite slab. A cold brick may be placed on each side to keep it straight. Then place a flat board across the top of the back. By this means your goose will assume a nice plump appearance by the time it is thoroughly cool and ready to be packed for the market. If, however, the goose is intended for consumption at home, it should be hung up by the legs in a cool place until it is required by the cook. This latter plan making the flesh more tender and a better colour. The next thing to be done before sending to market is the "packing." This is by no means the least important part of the business. Carelessness in packing has often resulted in undoing and ruining the work of months' previous industry. Hampers for packing geese or poultry of any kind for market should be strong and firm. They should be just the size required—neither too large nor too small—as either would be likely to bruise the skin and make them unsightly. As each layer of geese is placed in the hamper a piece of cheap calico or a sheet of clean white paper should be thrown over them, and all the spaces well filled up with sweet hay. Another sheet of paper or a piece of cotton should be placed on the top of the packing before the next layer of geese is put into the hamper. Every crevice should be well filled in with packing to guard against the possibility of shaking about in transit. The quickest route is certainly the best by which to send geese to market. They never improve on a journey. The sooner they can be conveyed from the homestead to their destination the greater will be the chance of obtaining a good price for them. The sooner they are unpacked the better they will look and greater will be the returns.

DISEASES.

I SHALL not attempt to deal with the cure of diseases in Ducks and Geese, but content myself with trying to show my readers how to prevent diseases from visiting their little flocks.

CRAMP.—Cramp in Ducklings is the most fatal of all diseases to which Waterfowl are subject. It is the forerunner of either death, deformity, or imperfection for life.—It is frequently caused by improper feeding and coddling, and by keeping them upon damp stone floors, or on the damp ground at night. Feeding on improper food, and being deprived of green food, grit, and water to swim in is often the cause of indigestion, which in many cases is the immediate cause of CRAMP.

The CROP of the Duck is intended by nature to prepare the food for the gizzard to grind, when thoroughly mixed with the ingredients which digest it, therefore it should be remembered that it is absolutely necessary, especially when using an extra quantity of soft food, to supply your birds with an abundance of grit and green food, to keep their digestive organs in proper working order.

Cramp is the most manifest symtom of other diseases.—The digestive organs once weakened, general debility follows upon the slightest provocation. Over crowding, damp floors, and improper food are the chief causes of cramp. Ducks and Geese, like most other birds and animals, are invigorated by fresh air and water. Their limbs and muscles must have exercise, and if a proper system of "Housing and Feeding" be carried out, there will be no Cramp.

Prevention is better than cure, and I would prefer setting another batch of eggs and rearing a fresh lot of ducks or geese than I would attempt to cure one single bird of cramp, for they are seldom good for anything if even they appear to be cured. Therefore I would advise the cause to be removed. Make their house dry and comfortable, especially the floor. Observe perfect cleanliness in all your arrangements, and let your ducks and geese have liberty, water, and exercise, then you will effect more than can be done by any medical treatment.

SOFT EGGS.—Waterfowl are sometimes troublesome during winter and early spring by laying soft eggs, which of course are of no use for incubation. To prevent these soft eggs, always place within easy reach of your birds an abundant supply of old mortar, burnt oyster shells, shell gravel, ashes, &c.; and if you have a stream, and the flow of water is not a very large one, throw a lump of fresh lime into it once or twice a week, or if you have a pond, put a few lumps of lime in it, and do *not* overfeed, and do not give them any animal food at all if you are troubled with soft eggs.

EGG-BOUND.—This is very rare in young birds, although it does occur sometimes. The best way to prevent this is to feed your birds well and wisely. Good sound food of a nutritious, but *not* a fatty, nature Let them have plenty of exercise.

GIDDINESS, or STAGGERS.—A recent writer tells use that a duckling's skull is so very thin that the sun seems to affect the brain very soon, and still he tells us that ducklings will do as well without water as with it for ten weeks. Such statements are enough to make us ask the question—Did that writer ever keep a duck? Or has some one told him that ducks will do as well without water as with it. If my readers wish to avoid giddiness, staggers, leg weakness, liver diseases, and many other ailments incidental in duck-keeping, they will do well to give their birds a reasonable supply of sound food, plenty of fresh air, exercise, and clean water to drink and swim in, together with a comfortable house to sleep in and a good bed of clean straw to lie upon. If you attend to my remarks in this article, vendors of medicine for ducks, etc., will not make a fortune, for there will be no medicine required.

Should you by accident have a bird cut or torn in any way, the wound should be dressed with Condy's Fluid or Carbolized Oil, and if necessary the wound may be sewn up with a fine needle and silk; but unless the bird is a valuable one, kill it in preference.

INDEX.

	PAGE
Preface	3
Introduction	4
Houses for Breeding Ducks	6
Duck Houses and Enclosures	9
Ponds	13
How to begin and Selection of Stock	15
Modern Aylesbury Ducks	21
Schedule for Judging do.	24
Modern Rouen Ducks	27
Colour of Drake (Rouen)	28
Do. Ducks	32
Schedule for Judging Rouen Drakes	33
Do. do. Ducks	35
Modern Pekin Ducks	37
Schedule for Judging ditto	40
The Cayuga Duck	43
Schedule for Judging ditto	46
Indian Runner	49
Schedule for Judging Indian Runner Ducks	54
Breeding and Treatment of Breeding Stock	56
Incubation	61
Breeding and Rearing Ducklings for Exhibition	67
Do. do. for Market	74
Method of Killing	77
How to Treat Ducks when Moulting	78
Preparing and Keeping Ducks in Condition for Exhibition	79
Trimming	83
Exhibiting	85
Transit	86
Penning	87
Cramming	88
The Question of Profits—1st Year	90
Do. 2nd Year	94
Do. 3rd Year	97
Geese	100
Do. Toulouse	105
Schedule for Judging ditto	109
Embden Geese	111
Schedule for Judging ditto	113
Breeding and Rearing Geese for Exhibition	114
Geese for Table and Market purposes	118
Diseases	124

ADVERTISEMENTS.

AYLESBURY DUCKS.

THIS Champion strain won outright, in three successive years, the first Ten-Guinea Challenge Cup ever offered for the above variety.

Young birds bred by me in 1894 won the highest honours at most of our English exhibitions, including the Ten-Guinea Challenge Cup at the Waterfowl Club Show, which was held at the Crystal Palace; first, Dairy Show, Liverpool, etc.

Ducks and Geese bred by me have won upwards of 2,000 cups, medals, and money prizes during the last ten years. Not only in England have my stock been successful, but in Australia, America, South Africa, New Zealand, and Newfoundland they have distinguished themselves and their breeder. Hundreds of unsolicited testimonials could be published in favour of stock and eggs from these yards. The excellency of my stock is so well known that further comment is unnecessary.

Eggs for Hatching, 21/- for Eleven.

EXHIBITION AND STOCK BIRDS FOR SALE
at prices according to quality.

HENRY DIGBY,
THE BURNE, BIRCHENCLIFFE,
HUDDERSFIELD.

ADVERTISEMENTS.

Indian Runner Ducks.

HENRY DIGBY,

THE BURNE, BIRCHENCLIFFE,
HUDDERSFIELD,

Has the best strain of Fawn and White "Indian Runner" Ducks in England.

Advertiser has only shown this variety once this year, (1897), viz.: at the Crystal Palace, where his Indian Runners carried off four firsts and two second prizes, also the cup for the best bird of this variety.

H. D. is also the breeder of the first and second prize Dairy winners, and also the breeder of the two first prize winners at Birmingham, 1897.

STOCK OR EXHIBITION BIRDS ALWAYS FOR SALE.

Eggs in Season, 10/6 for Eleven.

ADVERTISEMENTS.

EAST OF ENGLAND
Live Stock and Poultry Farms
THUXTON, NORFOLK,

Headquarters for High-class Exhibition and Breeding Pens of all Varieties of Prize Poultry, Ducks, Geese, Turkeys, and Bantams.

By far the Largest and Oldest Breeding Establishment in the World.

More Cups, Medals, Diplomas, and Prizes have been won by Birds from these Yards than any other two Yards put together. Our strains are noted for their wonderful Laying Qualities.

BIRDS OF ALL VARIETIES FOR SALE, FROM 7/6 EACH.

Eggs for Setting in Season, Guaranteed Fertile, and from same Birds as we breed from ourselves.

Before Purchasing elsewhere send for our Illustrated Descriptive Catalogue, containing List of Prizes and Testimony from Customers in all parts of the World. Free on application. Apply—

Abbot Brothers.

ADVERTISEMENTS.

WM. BYGOTT

Has always on hand

HIGH-CLASS ROUEN DUCKS

— AND —

TOULOUSE GEESE.

THE Cups, Specials, and Medals won annually by his ROUENS at all the leading exhibitions, including The Royal Agricultural Society of England's Shows, London Dairy Shows, Crystal Palace, and Birmingham, and the high prices of ten, fifteen, twenty, and twenty-five pounds each realized by a score of noted winners, including Challenge Cup Ducklings, is a sufficient proof of the intrinsic value of these strains.

W. B. can always supply Drakes and Ducks unrelated, as they are bred in several companies. Many noted breeders and exhibitors having had their stocks from these yards for several years now, with exceptionally good results, both as regards breeding and exhibiting.

RYEHILL FARM, ULCEBY,

LINCOLNSHIRE.

ADVERTISEMENTS.

CHAMPION WHITE LEGHORNS.

J. A. CHEETHAM,
OAKLANDS, BRIGHOUSE,
YORKSHIRE.

DURING the last ten years birds from this yard have been awarded the highest possible honours at all the best Shows. The following are a few of the awards:—

The White Leghorn Ten-Guinea Challenge Cups, eight times; Four cups, Crystal Palace, and medal three times for best Leghorn in the Show; Three cups, Dairy Show, and twice medal for best Leghorn; The Leghorn Fanciers' Club Gold Medal for best White Leghorn, three times; and Cups Birmingham, Liverpool, Bingley, Halifax, etc.; Gold, Silver, and Bronze Medals too numerous to mention.

High-class birds for Stock or Exhibition purposes always for Sale, and Eggs in Season.

Prices on application.

ADVERTISEMENTS.

JOSEPH PARTINGTON,

THE WOODLANDS,

LYTHAM,

BREEDER AND EXHIBITOR OF HIGH-CLASS

ROUEN DUCKS,

BUFF and WHITE COCHINS,

SINGLE and ROSE COMBED ORPINGTONS,

GOLD and SILVER POLANDS,

Has always for disposal birds of each variety, either for Exhibition or Stock purposes, at prices according to quality.

Rouen Ducks from these yards have had a most successful career during the last twelve years, having won the Cup at the Crystal Palace and Birmingham twelve years in succession, and three Ten-Guinea Challenge Cups outright; besides Cups, Medals, &c., too numerous to mention, at most of the leading Shows in the United Kingdom.

INSPECTION OF BIRDS AND YARDS INVITED BY APPOINTMENT.

ADVERTISEMENT.

MRS. F. DAVIS,

Breeder and Exhibitor of

HIGH CLASS

Pekin and Cayuga Ducks,

Has always birds for disposal at prices according to quality, either for stock or exhibition.

Birds from this yard have succeeded in winning the highest honours for the above varieties, including the Challenge Cup at Liverpool, Cups Crystal Palace, and First and Special Prizes too numerous to mention.

Address—

WOOLASHILL,

PERSHORE.

ADVERTISEMENTS.

VINCENT G. HUNTLEY,

INNOX MILLS,

TROWBRIDGE.

BREEDER AND EXHIBITOR OF

HIGH-CLASS

ROUEN DUCKS.

Has always for Sale birds of either sex, suitable for Exhibition or Stock purposes, at prices according to quality.

Eggs for Hatching in Season.

Foreign orders promptly attended to.

PRICES AND PARTICULARS ON APPLICATION.

ADVERTISEMENTS.

NINE FIRST PRIZES.
FIVE GOLD MEDALS,
SILVER & BRONZE MEDALS
HAVE BEEN AWARDED THESE

The Most Perfect Appliances for Hatching and Rearing.

THE WESTMERIA INCUBATOR

THE 108-EGG SIZE.

FOR 56, 108 & 216 EGGS.

Hatched at the Royal Show 89 per Cent.
,, ,, Dairy Show 93 per Cent.

"It Hatched all the Duck Eggs put in."—SIR FREDERICK MUSGRAVE, EDEN HALL, CUMBERLAND.

"I am more than satisfied."—HENRY DIGBY, ESQ., HUDDERSFIELD.

THE WESTMERIA BROODER

"No one having tried it will use any other."— THE DUCHESS OF WELLINGTON.

"The most perfect I know."—LADY PHILLIMORE, BOTLEY, HAMPSHIRE.

"By far the best. No Exhibitor should be without one."—G. H. PROCTOR, ESQ., DURHAM.

SEND PENNY STAMP FOR COMPLETE LISTS.

Westmeria Co., Leighton Buzzard,
ENGLAND.

ADVERTISEENTS.

BANTAMS.

MRS. ENTWISLE,

Breeder and Exhibitor of the

HIGHEST-CLASS BANTAMS,

Always has for Sale some Exhibition and Stock birds of the following varieties:—

GAME BANTAMS,

Black-Red, Brown-Red, Pile, Duckwing and Birchen-Grey.

PEKIN BANTAMS,

Buff, Partridge, White, and Black.

Also RED MALAYS and SEBRIGHTS.

Prices according to quality.

Special attention given to Foreign orders.

Birds can be seen by appointment.

Address—

Mr. J. F. ENTWISLE,

Calder Grove House,

Near WAKEFIELD.

"BANTAMS,"

By W. F. ENTWISLE.

THE RECOGNISED STANDARD WORK.

Illustrated by LUDLOW.

Crown 4-to., Bound Cloth, Gold Lettered.

6/- Post Free, from

MISS ENTWISLE,

CALDER GROVE HOUSE,

Near WAKEFIELD.

ADVERTISEMENTS.

CHAMPION INDIAN GAME.

WM. BRENT,
CLAMPET FARM, CALLINGTON,

CORNWALL,

Has always a large number of reliable birds for Show and Stock purposes on sale, and Eggs in season.

The strain is noted for their great size, shape, colour, and lacings, and 25 years careful breeding. Many thousands of prizes have been won by birds bred in these yards, at the best shows in the kingdom. (The owner himself having won Cups, Medals, and Money prizes with 62 *different birds*, from September 1896 to September 1897). Exports have been made U.S.A., Canada, New Zealand, Australia, Africa, &c.

SATISFACTION GUARANTEED.

The Rev. HAROLD BURTON,
FAULS VICARAGE,

PREES Near WHITCHURCH.

Breeder and Exhibitor of High-class

LIGHT BRAHMAS,

Has always for sale Stock or Exhibition birds at prices according to quality. The reputation of the birds annually bred in these yards is so well known that comment is unnecessary.

EGGS IN SEASON.

PRICES ON APPLICATION.

ADVERTISEMENTS.

ALWAYS ON SALE.
GOLD AND SILVER WYANDOTTES,
BARRED ROCKS,
ROSECOMB BANTAMS,
AND ALL VARIETIES OF HAMBURGHS.

Address—

H. PICKLES,
EARBY, COLNE.

ADVERTISEMENTS.

Pigeon Marking Conference Rings,

Registered No. 182427.

Price, 1/9 per doz.; 18/- per gross, consecutively numbered to 144.

Poultry Marking Conference Rings,

FOR EVERY VARIETY OF POULTRY.

Water-fowl Club Rings,

Pedigree Rings,

Registered No. 181730.

Suitable for every variety of Pigeon, Poultry, and Water-Fowl.

Send stamp for illustrated and descriptive price list.

Inventor and Sole Manufacturer—

HENRY ALLSOP,

89, SPENCER STREET,

BIRMINGHAM.

ADVERTISEMENTS.

Champion Brown Leghorns.

JOHN HURST,
SOUTH TERRACE,
GLOSSOP,

Breeder of ten Challenge Cup winners. The only winner outright of both Challenge Cups in three consecutive years. The Leghorn Fanciers' Club's Gold Medal (twice). Cups, Palace, Dairy, Birmingham, &c.

BIRDS FOR SALE.

JOHN WHARTON,
HONEYCOTT, HAWES, YORKSHIRE.

BREEDER OF BUFF, PARTRIDGE AND BUFF-LACED WYANDOTTES.

J. W. was the first to breed Buff Wyandottes in England, and has won highest honours at all the largest shows, including Firsts, Crystal Palace, Dairy, Birmingham, &c.

**Stock and Exhibition birds always on Sale.
Foreign Buyers liberally treated.**

ADVERTISEMENTS.

IF YOU WISH TO SUCCEED
IN
DUCK BREEDING,
USE
SPRATTS PATENT
POULTRY MEAL.

In Sealed Bags, and 3d. and 4d. Sample Packets. Per cwt. 20s., per half-cwt. 10s. 6d., per quarter cwt. 5s. 6d., per 14-lb. 2s. 9d., per 7-lb. 1s 6d.

Give during Severe Weather increased proportion of

Granulated PRAIRIE **Crissel** MEAT

A Preparation of Meat taking the place of Insect Food.
See you get it in Sealed Bags.

Per cwt. 26s., per half-cwt. 13s. 6d., per quarter-cwt. 7s., per 14-lb. 3s. 9d., per 7-lb. 1s. 11d. Samples post free.

THE COMMON-SENSE OF POULTRY KEEPING
3d. Post Free.

Contains full and practical information on Poultry Rearing, Housing, Feeding, and Diseases

OF ALL DEALERS, OR OF
SPRATTS PATENT LIMITED,
BERMONDSEY, LONDON, S.E.

ADVERTISEMENTS.

SPRATTS PATENT
LIMITED,
MANUFACTURERS, BERMONDSEY,
LONDON, S.E.

Baskets for Ducks and Geese made of the best White Wicker-work.

UPRIGHT SHOW BASKETS.

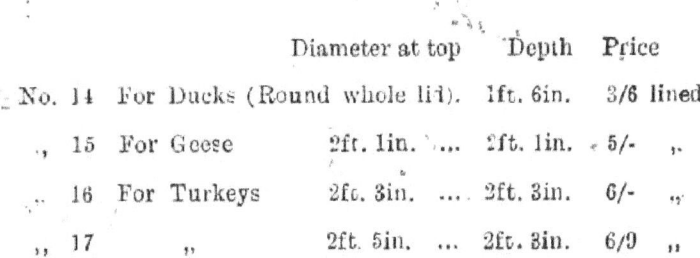

		Diameter at top	Depth	Price
No. 14	For Ducks (Round whole lid).		1ft. 6in.	3/6 lined
,, 15	For Geese	2ft. 1in.	2ft. 1in.	5/- ,,
,, 16	For Turkeys	2ft. 3in.	2ft. 3in.	6/- ,,
,, 17	,,	2ft. 5in.	2ft. 3in.	6/9 ,,

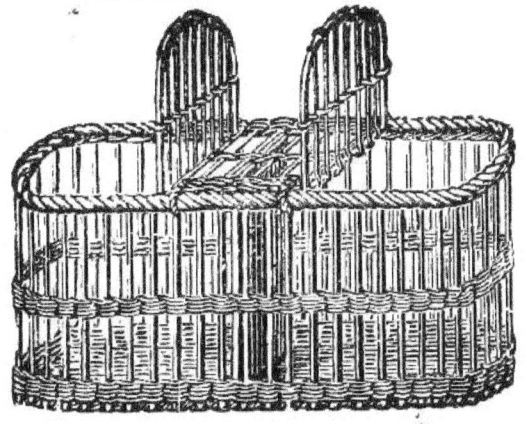

DOUBLE SHOW BASKETS.

| No. 22 | For Ducks, lined | | ... | 12/6 |
| ,, 23 | ,, | ,, | larger | 14/6 |

MANUFACTURERS OF

Duck Houses, Troughs, Baths, &c., also Poultry Houses, Pigeon Cotes, Aviaries, and every Appliance for the Poultry Yard and Kennel.

ILLUSTRATED PRICE LISTS, POST FREE.

www.ingramcontent.com/pod-product-compliance
Lightning Source LLC
Chambersburg PA
CBHW062322220526
45469CB00008B/2599